BUTTERFLY CONSERVATION

BUTTERFLY CONSERVATION

T. R. NEW

Second Edition

Melbourne
OXFORD UNIVERSITY PRESS
Oxford Auckland New York

OXFORD UNIVERSITY PRESS AUSTRALIA

Oxford New York
Athens Auckland Bangkok Bombay
Calcutta Cape Town Dar es Salaam Delhi
Florence Hong Kong Istanbul Karachi
Kuala Lumpur Madras Madrid Melbourne
Mexico City Nairobi Paris Port Moresby
Singapore Taipei Tokyo Toronto

and associated companies in
Berlin Ibadan

OXFORD is a trade mark of Oxford University Press

First edition published 1991
This edition published 1997

National Library of Australia
Cataloguing-in-Publication data:

New, T. R. (Timothy Richard), 1943–.
Butterfly conservation.

2nd ed.
Bibliography
Includes index.
ISBN 0 19 554124 3.

1. Biological diversity conservation. 2. Wildlife conservation. 3. Butterflies — Ecology. I. Title.

333.955716

Text and cover designed by Anitra Blackford
Typeset by Syarikat Seng Teik Sdn. Bhd., Malaysia
Printed through Bookpac Production Services, Singapore
Published by Oxford University Press,
253 Normanby Road, South Melbourne, Australia

CONTENTS

ACKNOWLEDGEMENTS

The following are thanked for permission to reproduce or modify illustrations from published papers or books: Academic Press, Orlando, Florida; Dr M. J. Adams; Bailliere Tindall; Blackwell Scientific Ltd; Butterfly Conservation; Chapman & Hall Ltd; Dr S. P. Courtney; Professor W. H. Dowdeswell; Elsevier Science Publishers Ltd; International Union for the Conservation of Nature and Natural Resources (The World Conservation Union); Dr M. Ishii; Kluwer Academic Publishers; Lepidoptera Research Foundation; Lepidopterological Society of Japan; Manchester University Press; The Natural History Museum, London; New York Zoological Society; Outdoor California; Penguin Books Ltd; Dr E. Pollard; Royal Entomological Society; Society for Conservation, Biology; Springer-Verlag, Heidelberg; Taylor & Francis Ltd; Dr M. S. Warren.

I also gratefully acknowledge the advice I have received in response to specific queries from colleagues in many parts of the world. The topic of Lepidoptera conservation is a dynamic and rapidly advancing one, and the material cited in this book is based on published work or, in a few cases where I have received specific permission, refers to unpublished data: these are noted as 'pers. comm.' in the text.

Some of the illustrations were adapted by Mrs Tracey Carpenter, who also typed much of the first draft of both editions. Editorial comments from Ms Bronwyn Collie (first edition) and Dr David Meagher (second edition) improved the manuscript substantially.

PREFACE TO THE FIRST EDITION

There are more books on butterflies than on any other group of insects. They fall, very broadly, into two categories — those which are essentially identification guides to, or synoptic treatments of, the butterfly faunas of a given geographical area, and those which are more general treatments of their natural history or, more formally, 'biology'. Their abundance is testimony to the interest which many people have in butterflies. In addition, there is a vast technical literature of research papers and reports which are usually not as easily available to the general reader and which continues to burgeon as more and more naturalists and scientists contribute their observations and thoughts on butterflies in many parts of the world. Many people picking up this book are likely to already have several others on butterflies, and may think twice before buying another. Indeed, whilst planning the scope of this work, I was relieved to read in the preface to *Butterflies of the World* (Preston-Mafham and Preston-Mafham, 1988) the following, 'Our thoughts inevitably turned to bookshop shelves already creaking under the weight of butterfly books ... At first sight it appeared that we would find it difficult to present anything original in any work which we could possibly contemplate writing'. This doubt summarized my own feelings quite adequately. So, why yet another book on butterflies?

Many people are becoming more concerned about the wellbeing of our natural world as our intrusions into natural ecosystems change (often irrevocably) their suitability for the animals and plants which live there. Conserving large or popular animals and rare plants is now readily accepted as a worthwhile, even necessary, aspect of ecosystem management and arouses widespread public sympathy. Yet suggestions of conserving the multitude of smaller or less conspicuous animals, invertebrates, in our world still commonly meet with suspicion, incomprehension or scepticism. Butterflies have received a great deal of attention during this century, and before, and suggestions of conserving them are often taken more 'seriously' than similar promotion of other kinds of insects. Thus they have a vital role as ambassadors for a greater conservation need, and the lessons being learned from their conservation during recent decades have been the greatest single factor in transforming the topic of invertebrate conservation from fantasy to an increasingly sophisticated science, the principles of which are of much wider relevance to land use planners and conservation biologists than might initially be apparent. For some such people, this might even be their first book on butterflies as they seek to broaden interests from other fields of biology.

I have tried to summarize much of the highly scattered information which is available on the conservation of butterflies themselves, and on their roles in conservation assessment. The first chapter introduces some of the rationale of insect conservation and indicates how butterflies fit into this. The second is a broad introduction to butterfly classification and biology, stressing the individuality of each of the 20 000 or so different species and summarizing the kinds of information needed for practical conservation. The remaining seven chapters explore aspects of butterfly conservation more directly, ranging from national or international efforts to safeguard species or communities to the steps that we can take as individuals to help encourage butterflies, and document their biology in our own gardens and neighbourhoods. They therefore help to answer the question 'What can *I* do?', and to place this in a broader and informed context. I have assumed very little previous knowledge by the reader, and a short glossary of some technical terms will, I hope, help to overcome my lapses in this intention. But I have also incorporated some detailed recent work, and felt it useful to document reasonably fully much of what I have written, and the bibliography should be of use not only to the less experienced lepidopterist but also to those who wish to read more extensively about various topics and cases discussed in order to obtain more background to help with the practicalities of other, related, situations. I have included information available to me until late 1989.

It will soon become apparent that most work has been done in Europe and the United States of America. In many other parts of the world advances are indeed being made, and the political climate for invertebrate conservation is now much more sympathetic than it was a very few years ago.

This work had its gestation in a booklet I wrote in 1987 which was sold to help the 'Eltham Copper Appeal' (pp. 182–4). That work aroused interest from many people to whom butterfly conservation was a novel, or even alien idea, and several readers suggested that it should be expanded and updated to form a source book of information and ideas on this topic. Although Australian examples are included where possible, the book is of much wider scope. 'Butterfly conservation' now almost always merits a mention, or even a chapter, in books on butterfly biology, but it is not the predominant theme of any other volume written for a global audience. Several European and North American books contain substantial data on conservation of those regional faunas, and I hope that this broader treatment may complement these and increase their relevance elsewhere.

The illustrations of butterflies in this book* have been selected from a number of early works on these insects, to exemplify not only a range of different kinds of butterfly but also the skill and appreciation with which they have been depicted over many years; again, evidence of the special place they have held in the esteem of many artists and naturalists over a very considerable period.

T. R. New
Melbourne, March 1991

* Most not included in second edition

PREFACE TO THE SECOND EDITION

Butterfly conservation has made considerable progress during the last few years, as awareness of the enormous variety of organisms present on Earth increases, and the need to sustain the ecosystems on which they depend becomes more widely appreciated. Needs to document our natural world, and to seek groups of animals which can be used as 'indicators' (to reflect the condition of the environment) or 'predictors' (as surrogates of wider 'biodiversity') are urgent in practical conservation assessment, and local or national awareness of the decline and vulnerability of butterflies has led to an accelerating number of species which are the focus of specific conservation programmes.

With the realization that detailed autecological studies are the key to success, rather than simple reserve declaration or legal protection alone, resources available for butterfly conservation have increased (though, by no means enough!), and our understanding of the dynamics of butterfly populations and assemblages has made substantial advance.

Importantly, many people — not simply a small group of specialists — are discussing the core topics and strategies at many levels. The recent expansion of the UK organization Butterfly Conservation to nearly 10 000 members is unlikely to be paralleled in many less-developed parts of the world, but butterflies appear at some level on the conservation agendas of very many countries, and the strategy of 'conservation through development' is progressively incorporating them. This book refers to papers published in the proceedings of several major national and international meetings on Butterfly Conservation held in the last few years: with very few exceptions, such meetings would have been regarded as eccentric only a couple of decades ago. As Pyle (1995) noted in his historical review of Lepidoptera conservation, 'In just a quarter of a century, Lepidoptera conservation has grown from an arcane topic to a commonplace concern.'

The central place of butterflies in promoting conservation awareness of other invertebrate animals is reflected strongly in recent texts on insect conservation (Samways, 1994) and invertebrate conservation (New, 1995), both of which contain numerous references which expand the scope of this book.

The general structure and content of the first edition has been retained, together with the aim of providing a broad biological background to butterflies. I have added sections dealing with recent developments — such as the impact of the 'metapopulation concept' on understanding butterfly populations — and have included about 150 more recent references, reflecting information available to me up to September 1996.

T. R. New
Melbourne, January 1997

CHAPTER ONE

INTRODUCTION

Butterflies are insects, representatives of that most abundant and diverse group of animals which dominates virtually all of the world's terrestrial and freshwater environments. But they are unusual insects in that they generally appeal to people, and we still hear talk of 'butterflies and insects' as if these are different groups of animals. Watching and studying butterflies, evocatively discussed by Pyle (1992), perhaps ranks second only to ornithology as a pastime for naturalists interested in animal life. However, many people, including those strongly committed to the wellbeing of our natural world, find the idea of conserving insects and other kinds of small animals alien or, at least, strange. We have been conditioned to varying extents by advertising and other media exposure towards the belief that such creepy-crawlies are bad, and should be eliminated; the chemical pesticide industry is large and influential, and the need to protect agricultural, horticultural, orchard and forestry crops, stored grains, fruits and other foodstuffs, clothes and construction materials from insect attack generates the largest employment field for scientists who study insects, entomologists. Thus, the suggestion of conservation — rather than eradication or suppression — of animals such as flies, beetles or cockroaches, together with other invertebrates such as spiders, earthworms or snails, can come as rather a shock. A mention of conserving butterflies, in contrast, can often generate immediate sympathy.

Consider our common reactions to the terms 'butterfly' and 'moth' — insects belonging to the same order (Lepidoptera) and sometimes so similar that only a specialist may be able to distinguish them with confidence. The very word 'butterfly' implies colour and motion. It tends to foster the images of warm sunny weather, tranquility, colourful etherealism and 'all's well with the world'. In contrast 'moth' commonly invokes the idea of furtive nocturnal creatures which get in one's hair, eat one's clothes or crops and are pests — definitely second-class animals judging by much public reaction. A few butterflies *are* pests — the European Cabbage White, introduced into Australia about 1940, is one example and, interestingly, many gardeners in Victoria (where it can cause severe depredations of cabbages and other *Brassica* vegetables) habitually refer to it as the 'cabbage moth'. Caterpillars of some skipper butterflies feed on grasses or palms, and those of a few larger nymphalid (see next chapter) butterflies in the tropics are pests of bananas. A few other butterflies are occasionally pests of citrus (some swallowtails) or legume crops (some blues). These tend to be ignored in areas where they do not cause major depredations. In contrast, most moths are also innocuous but tend to have their reputations tarnished by the sins of the minority.

Unlike most other insects, even their closest relatives, butterflies are accepted readily as being aesthetically and culturally desirable, and as intrinsically worthy of protection when the need arises. They are the insect equivalents of the Blue Whale, Giant Panda, Rhinoceros and Californian Condor. This book is about their biology and wellbeing, and the place they hold as a 'flagship' group of lower animals in the study and practice of invertebrate conservation, at a time when many species have declined markedly in abundance, or become extinct, as a consequence of human activities: the various endangering processes to butterflies are discussed in Chapter 3. Many butterflies *need* active conservation if they are to persist far into the future. This first chapter provides some general background rationale to the theme of insect conservation and indicates why butterflies are important in its pursuit.

CONSERVING INVERTEBRATES

The widespread conceptual barriers to conserving lower animals are gradually being overcome, and many people now admit their importance in natural ecosystems and in maintaining our natural world. Much of the reasoning behind this change is summarized in a document termed the 'European Charter on Invertebrates', which was adopted by the Council of Europe, Strasbourg, in 1986; a slightly modified version was promulgated in Australia in 1989. The Charter sets out many of the reasons why invertebrates (all those animals without backbones, in contrast to 'vertebrates' — the fish, amphibians, reptiles, birds and mammals) merit positive attention. Collectively, these constitute an overwhelming case for their conservation: in anthropocentric terms there are quite as many 'goodies' as 'baddies', and the great majority do us no harm and do not impinge directly on our attention.

The major points from the Charter are as follows, with some of the other points included in that pioneering document:

- Invertebrates are the most important component of world fauna, both in numbers of species and biomass. We simply do not know how many different kinds (species) of animals exist but, whereas there are probably of the order of 50 000 vertebrate species, estimates of the numbers of arthropods alone (the large group of invertebrates including the insects, together with spiders, mites, centipedes, crabs, barnacles and similar beasts with jointed limbs and a hard outer casing) run into several tens of millions. About a million species of insects have been formally recognized by scientists, and thousands more are being described each year. In Australia, probably less than half of our native insects have yet been dignified by the basic step of being named.

  The total numbers of invertebrate individuals, and their biomass (amount of living matter), clearly exceed those of vertebrates. In Europe, soil invertebrates may reach around a tonne for each hectare of ground area, and flying insects above the ground can constitute more than 100 kilograms per hectare. In another major environment, the sea, invertebrates such as molluscs and crustaceans are fished collectively at a rate of 9–10 million tonnes each year. Together, they and other invertebrates play vital roles in recycling materials so that organic matter 'tied up' as plant and animal structure remains usable in ecosystems. There is little doubt, despite the relative inconspicuousness of many kinds, that insects and other invertebrates are the most important animals in maintaining the natural basic functions of ecosystems and communities.
- Invertebrates are an important source of food for animals. Many birds, amphibians, fish and other vertebrates feed wholly or largely on invertebrates, and large groups of mammals (such as bats and the aptly-named insectivores) also depend on them for food. The largest of all mammals, the great whales, feed almost entirely on marine crustaceans, krill. Clearly, damage to such specific food materials also has far-reaching effects on the consumers and the communities of which they form a part.
- Invertebrates may also constitute an important source of food for people. When we hear politicians or other people influential in making decisions on conservation deriding or denigrating the importance of invertebrates, a rapid change of heart can sometimes be induced by asking if they eat honey, or like oysters, prawns or crayfish! Bees, molluscs and crustaceans are widely utilized in food industries, and sustain considerable commercial activity. Less well-known, but just as important as foodstuffs to many people, are termites, grasshoppers, insect larvae (such as witchetty grubs) and spiders.
- Invertebrates are vital to the fertility and fertilization of most cultivated plants. Most of the plants utilized by humanity are pollinated by insect vectors, often ones quite specific to particular colours or shapes of flowers.

Bees are the most commonly perceived pollinators, but many other insects — including butterflies — are also important. The wide-ranging effects of soil invertebrates such as earthworms essentially regulate the fertility and productivity of vast areas of land utilized for crops. Many of the pests which infest desirable and useful plants are themselves destroyed, or kept from large population increases, by other invertebrates: various insect predators and parasites attack almost all plant-feeding insects and play major roles as natural biological control agents. Many have been manipulated by pest managers to augment their protective effects on crops.

- Invertebrates are valuable aids in medicine, industry and crafts. Many scientists believe that the role of these animals as sources of medical preparations has hardly been explored, and some traditional uses are of considerable age — the use of leeches, and the use of bee stings to relieve symptoms of rheumatism and arthritis, for example. Industries such as sericulture (silk from the cocoons of moths, predominantly of the 'silkworm' *Bombyx mori*), pearls and corals for jewellery, sponges, shell ornaments, and others, may constitute vital facets of local community life and be major export industries in places.
- Many invertebrates are of considerable aesthetic value. The European Charter also points out the important roles that invertebrates may play in research in biology, medicine, chemistry and other fields. As just two examples, the small fruitfly (or vinegar fly) *Drosophila* has been highly significant in clarifying the principles of heredity, and research on breeding swallowtail butterflies elucidated the highly significant medical problem of rhesus factor inheritance in humans and allowed steps to be taken to counter potential lethal effects.

The practical reasons for conserving invertebrates noted above are persuasive, substantial and incontrovertible, yet many people would argue that such anthropocentric pragmatism — conserving things because they are useful or 'good' for us, thus having 'utilitarian' values — should take a subordinate role to their intrinsic 'right to exist' (Lockwood, 1987). Unfortunately, such arguments, however fully we may advocate them, are often *not* as persuasive in a political context, and the conservation argument usually still needs to be couched in more practical terms. Whether or not the views of some schools of 'animal rights' advocates are acceptable is not the major point here — ethical ideals are clearly a vital facet of conservation concern — but short-term practical considerations and human interests dominate much argument about ecosystem management. However, many people do find butterflies intrinsically worthy of conservation, perhaps largely because they are aesthetically 'pleasing': although there are several explanations of the derivation of the name, it has been suggested that the very word 'butterfly' may be a corruption of 'beauty-fly'.

Butterflies are also perceived as being harmless. They do not bite or sting, and most species are silent rather than 'threatening' us with buzzing or other noises.

Butterflies have a long history of aesthetic appeal. Recognizable butterflies are depicted in ancient Egyptian paintings and in mediaeval illuminated manuscripts as well as in a host of later art. Their overall 'charisma' (a term which can be applied widely to very few groups of invertebrates; perhaps dragonflies and some molluscs are the other major contenders) has ensured that they have received considerable attention in many parts of the world, initially by naturalist-collectors seeking to finance their explorations in remote parts of the world by selling specimens to wealthy patrons and other collectors. Their aesthetic role has been augmented culturally by butterflies being considered as symbols of the Resurrection (the rebirth of caterpillars in a different form when they emerge from the intermediate pupal stage) in some Christian societies. There are also early account of 'rains of blood' which are attributable to butterflies. The Reverend William Bingley in his *Animal Biography* (1804) claimed that this was first recorded in France in 1608, when a shower of 'blood' over several miles around Aix caused considerable alarm. An explanation was provided by the philosopher de Peiresc, who happened to pick up a large butterfly pupa and keep it in a box; on emergence to the butterfly, it left a drop of blood-coloured liquid (meconium), and de Peiresc concluded that the shower of blood represented the mass emergence of vast numbers of butterflies during the same short period. However, Bingley had overlooked at least one earlier record of this phenomenon: Mouffet (1658) referred to a case in Germany, in 1553 where 'an infinite army of Butterflies . . . did infest the grass, herbs, trees, houses and garments of men with bloody drops, as though it had rained blood'.

From around the middle of the nineteenth century up to the present day, butterfly collecting has been an absorbing and instructive hobby for myriads of people. There are possible conflicts between collecting and conservation, which we will examine later, but a major result of this tradition is that butterflies are, at least superficially, amongst the best-understood groups of insects and one of the few groups not the near-exclusive province of professional scientists. Such fundamental knowledge is of immense value in planning conservation and documenting natural communities, and it is worthwhile to compare butterflies with most other groups of insects from this point of view.

I noted earlier that there are a formidable number of different kinds of insects. Many text books relate figures of 2–3 million species as a minimum, but these estimates are based on projections from relatively simple collecting techniques. Some vastly increased estimates have resulted from recent surveys of tropical forest canopy, hitherto an almost inaccessible habitat for biologists to study and one which is proving to support enormous numbers of insects which have not been found elsewhere. Invertebrates living in the tops of trees can now be sampled by using non-persistent pyrethrin insecticide mists from power sprayers hoisted into the canopy on ropes and catching the falling insects on plastic trays or in large funnels suspended near the ground. Extrapolation from catches of beetles on selected trees in South America to encompass the enormous diversity of tree species (each with specific plant-feeding insects) in Amazonian forests led Erwin (1982) to produce an estimate of 30 million arboreal arthropod species in the neotropical region alone. Many people felt that this estimate was far too high, but other studies in south-east Asia suggest that, on a global scale, we may well have to enlarge even this impressive total. With the research resources at present available in the world's leading museums and tertiary institutions it is totally impracticable to document (that is, formally record and name) all the species of insects present, and this situation is unlikely to improve to any great extent. It is indeed sobering to reflect, as E. O. Wilson has done, that we know more about the number of stars in the universe than we do about the number of species that share our world.

This leads to what Taylor (1976) has aptly termed the 'taxonomic impediment' in entomology. We cannot, and almost certainly will never be able to, provide complete lists of species of the insects present in any major area, especially of the tropics, of the sort taken for granted by people studying mammals, birds, vascular plants, and others — the sort of 'inventory' on which much conservation assessment and management has traditionally been based through helping to interpret patterns of species diversity. An important ramification of this is that people who may influence conservation procedures and priorities, such as politicians and land managers, commonly take the lack of definitive species lists of invertebrates as symptomatic of disinterest by biologists, or lack of importance, rather than reflecting major ecological complexities. It means also that, with some exceptions, we cannot state categorically whether or not a particular invertebrate species is rare or otherwise worthy of conservation, because we do not know where else it occurs and what its detailed environmental needs may be. The taxonomic impediment thereby leads directly to a conservation impediment: lack of precise documentation weakens advocacy for conservation need. Even when, rather rarely, we can name most of the more obvious taxa present, we may know very little about their biology and how the various species interact with each other. As a simple example, we cannot at present provide complete documentation of the insect fauna of any given species of *Eucalyptus* or *Acacia* in Australia, let alone the broader community of which these are a part.

There are other practical problems with trying to survey invertebrates (New, 1984, 1995), one of which is worth stressing here. Unlike the majority of vascular plants, and many mammals and birds, many invertebrates may only be visible for a short time each year because of their brief adult life. They also occur in structurally and biologically different growth forms. Whereas a fledgling Eastern Rosella or a young Leadbeater's Possum grossly resembles the adult of the same species, and is thus readily associated with it and recognized, a beetle larva or a caterpillar is a very different biological entity from the adult. It may occur in a different habitat, utilize different foods, need different techniques to collect it and a different scheme of classification to recognize it — even if the juvenile and adult can be associated. Many cannot be identified, as the larvae have not been reared through in captivity to the adult stage, and this step is obviously needed to confirm the relationship. Many adult insects are active for only a few days or weeks each year, with the rest of the time being passed in inconspicuous or inactive growth stages. The species may therefore not be easily discovered during surveys at many times of the year. In fact, solving the difficulties of executing an invertebrate inventory to any reasonably comprehensive level within a reasonable period requires much original research and practical logistic support (New, 1996). Many practitioners seek 'short cuts' to facilitate getting useful biological information without the need for this major comprehensive understanding.

Butterflies are used commonly in this context of assessing terrestrial communities. Long-term collector interest has led to them being amongst the best-known insects; early stages of many species (particularly in northern temperate regions) have been described and associated with the corresponding adults and, probably, more has been written on butterflies than on any other group of insects. They are not particularly diverse in terms of invertebrates — somewhat fewer than 20 000 species have been described worldwide — and many countries have small and definable faunas. Australia has slightly more than 400 species and the UK under 60 species, for example. Not all species of butterflies have been named, of course, and several hundred species probably remain undescribed, but it is rather unusual to find undescribed large butterflies which are completely new to science, in contrast to groups of lesser-known insects such as barklice or lacewings in which it is still uncommon to find a majority of named species in some collections from the tropics! A new species of the butterfly genus *Idea*, with forewing length of 86 millimetres, was described from Sulawesi only in 1981, but many other descriptions of new larger butterflies tend to come from revisions of status of known taxa, such as upgrading of putative subspecies (see Chapter 2) to full species, or division of species when new sets of characters are studied. In contrast to groups such as beetles and flies (each with several hundred thousand species, many of which are not clearly recognizable), or even moths, it is thus feasible both to assess the butterfly fauna of a given region or habitat, and also to derive relatively accurate comparisons of the butterflies of different areas — perhaps as an aid to 'ranking' these as priorities for nature reserves or other forms of protection. Good identification guides, with coloured plates of all (or most) species are available for butterflies of many parts of the world, so that most kinds can be recognized by non-specialist field workers without recourse to more specialized examination. In Europe, particularly, early stages are also recognizable with reasonable confidence and much biological information is available, including data on how particular butterflies may respond to changes in land management involving human activities such as agriculture or forestry. Finally, observations on butterflies are facilitated by their being diurnal, with many tending to be most conspicuous on warm, sunny days — ideal conditions for ecologists to work in!

The relatively precise and restricted environmental requirements of particular butterflies means that these can have considerable value as 'indicator taxa' — groups which can shortcut total community documentation by furnishing information which indicates the broader effects of environmental change or reflects a particular suite of ecological conditions. Partly because of this, coupled with their general conspicuousness and ease of recognition, butterflies have figured very highly in insect conservation work. Public goodwill is an invaluable bonus to their use.

RATIONALE OF BUTTERFLY CONSERVATION

Two main themes recur in considering butterfly conservation, and both are treated more fully later on. The first concerns conservation of butterflies *per se*. Many butterflies are now far less abundant or widely distributed than they used to be, and declines of some have been precisely documented, with the times of their loss from particular sites known. As with other groups of animals, many of these declines can be attributed directly to human influences. The accumulated wisdom of collectors over a considerable period ensures that many naturalists are both aware of such trends and concerned about them. Particular species or populations of butterflies are thus accepted with little question as the prime targets of conservation activities, in the twin contexts of 'crisis management' for those which are threatened by immediate environmental change (such as destruction of their habitat) and prevention of a more gradual decline.

The second main theme concerns the use of butterflies to indicate community or habitat 'health'. It is often necessary, as noted earlier, to obtain comparative information on a suite of habitats in order to assess their suitability or relative priority for reservation or management, and evaluating areas in this way is a highly complex task (see, for example, a recent book on this theme by Usher, 1986). Among the many criteria which can be compared are diversity, presence of rare species, 'naturalness', proximity of threats, the area of habitat, and scientific value. For invertebrates, Disney (1986) suggested that our general ignorance precludes the use of many parameters available for studies on vertebrates and that — even in a well-known area such as the UK — only diversity, rarity, area and naturalness are really worth considering, and that all of these have their shortcomings in terms of objectivity unless documentation is relatively complete.

Butterflies are among the few groups of invertebrates for which relatively complete data may be obtainable, and they fulfil many of the conventional criteria stated to define useful indicator groups: day-flying and conspicuous, taxonomically tractable with most species recognizable, widespread, relatively diverse (but not too diverse for relatively complete assessment), many with precise ecological requirements and known to respond to particular changes in habitat parameters. One possible disadvantage is that most species are associated with plants so that, as a group, butterflies do not participate in as wide a range of ecological interactions as some other invertebrates. But they are undoubtedly useful in assessing habitats on a local scale (such as comparing a series of swamps, heaths, grasslands or woodlands in the same State or county) and on a broader geographical scale, where it is possible to detect 'critical faunas' or 'critical habitats' — areas of unusually high scientific significance which may be key environments for clarifying our understanding of evolutionary processes. Wilcox *et al.* (1986) concluded that butterflies (in parts of western USA) are at least as useful as vertebrate animals in planning conservation reserves. They are generally more apt to 'recognize' fine divisions of habitat, for example, and the scarcity of many vertebrates means that these may not be used as easily. Changes in butterfly status over time often indicate the broader effects of human disturbance to natural communities. Hammond and McCorkle (1984) regarded *Speyeria* butterflies (Nymphalidae) and their *Viola* food plants as being among the best indicator organisms of native undisturbed communities in North America. They are among the first taxa to be eliminated by disturbance, so that many local forms of *Speyeria* have declined rapidly over the last 200 years. Several have become extinct, or are threatened with imminent extinction, and *Speyeria* are by no means alone in being able to indicate more widespread changes to natural communities.

It is interesting to compare these practicalities with the butterflies' close relatives, moths. Apart from the general lack of public sympathy mentioned earlier (and there are important exceptions to this: the spectacular giant silk-moths and larger hawk moths, as well as some smaller brightly coloured day-flying groups such as the burnets, have long attracted collector interest and have been dubbed 'honorary butterflies'!) moths tend to be less conspicuous. They are considerably more diverse than butterflies, and vast numbers of species remain undescribed or are recognizable only with difficulty. Very little is known of the detailed biology of most non-pest species outside limited areas of the northern hemisphere and, whereas it is known that many species are indeed restricted in their ecological needs and geographical distribution, the available data base is both less complete and therefore less useful for more general conservation activities. We may be able to state with some confidence that a given area of Australia supports, say, 25 species of butterflies, six of which are rare, two of which are likely to decline in abundance if a given change is promulgated;

and that we know the gross needs of 20 of the species. Comparatively, the same area might include 250 species of moths, with us knowing very little about the distribution or biology of at least 245 of these. Moths could thus augment a case for conservation of an area in terms of diversity (numbers of species present), but perhaps not at other meaningful levels of interpretation or extrapolation because of the relative lack of ecological information. Certainly, some rare or remarkable species of moths have been the targets of species-orientated conservation measures in various part of the world (as have examples of many other groups of insects), but these are exceptional.

Essentially, then, butterflies are paramount among insects in invoking public interest and sympathy for their own wellbeing, and as valuable tools in the related contexts of developing methodologies for invertebrate conservation and in promoting both its practical and intrinsic worth. A comment by the famous nineteenth-century naturalist H. W. Bates in his *The Naturalist on the Amazons* (1864) (and used by Scudder, 1889a, and Holland, 1898, as a prefatory comment to their books, and quoted also by Brown, 1987) is proving to be prophetic as insect conservation makes its steady progress towards becoming an intricate and vital science: 'therefore, the study of butterflies — creatures selected as the types of airiness and frivolity — instead of being despised, will some day be valued as one of the most important branches of Biological Science'.

CHAPTER TWO

BUTTERFLIES — CLASSIFICATION, DIVERSITY, AND BIOLOGY

Butterflies comprise the smaller part, perhaps no more than about 10 per cent, of the insect order Lepidoptera (butterflies and moths), themselves closely related to the caddisflies or Trichoptera but distinguished by usually having their wings covered in flattened scales (rather than simple hairs) and the adults having a long coiled tongue or 'proboscis' for taking nectar from flowers. The term 'butterfly' is generally applied to members of two major groups, or superfamilies, the Hesperioidea (skippers) and Papilionoidea (all other butterflies) often collectively called the Rhopalocera, a term which is essentially synonymous with 'butterflies'. Members of a third superfamily, Hedyloidea, are now also considered to be butterflies, following a reappraisal by Scoble (1986), but all other Lepidoptera are conventionally treated as 'moths'. However, it is difficult to give an unambiguous definition of 'butterfly' and this problem has been aggravated by the inclusion of the Hedyloidea (formerly included in a large group of moths, the Geometroidea), as the assemblage of 150 000–180 000 or so moths is incredibly diverse in appearance, and many have one or more traditional 'butterfly features'. These features (and myths) include the following (partly after Feltwell 1986):

- Butterflies are active during the day and moths are active at night.
  Very few butterflies are nocturnal, although a few skippers, hairstreaks and nymphalids are attracted to lights, but many moths fly during the day time.
- The antennae of butterflies terminate in a well-defined knob or 'club', whereas those of moths do not.
  Butterflies never have really feathery antennae, but the knob may not be well defined and represented only by a very gradual or preapical thickening,

especially in some skippers. A number of moth groups, including some bright diurnal forms, have antennae which are swollen towards the tip.

- Butterflies are brightly coloured, moths are drab.
  Many butterflies are predominantly brown or greyish and are well-camouflaged when resting, whereas many moths are extremely brightly patterned in vivid reds, yellows, black, white or iridescent blue, often in spectacular combinations. In general, many day-flying moths are brightly coloured, and the colours may play a part in courtship or protection against predators. Those which are nocturnal tend to be less highly coloured, and so are much more cryptic during the long periods of daylight when they are at rest on tree trunks or other vegetation.
- When at rest, butterflies have their wings closed vertically above their bodies. Moths commonly extend their wings horizontally along their bodies or to the side, a trend which helps to reduce shadows and may also help to make them inconspicuous to predators. There are exceptions to both these generalizations — in the case of butterflies, many skippers and some Lycaenidae habitually extend their wings broadly.

There are other putative general differences, including some of a rather technical nature (such as the method by which the forewing and hindwing are held together in flight), but the above are sufficient to indicate the partial artificiality of the two groups. The partitioning is further confused by the Hedyloidea, a South American group of about 40 species, which have threadlike or slightly feathery antennae and are apparently mainly nocturnal. Virtually nothing is known of their biology (Scoble, 1986), and the life history of only one species has been described. They are therefore not considered further here.

CLASSIFICATION

I claimed earlier that butterflies are 'taxonomically tractable', so it may seem anomalous to admit now that it is difficult or impossible to state how many species there are! Several estimates have been made, ranging from a low of 13 000 (Owen, 1971a) to a maximum of 20 000 (Vane-Wright, 1978). Robbins (1982), in a compilation by major faunal regions, produced figures of 15 900–18 225 species. More recently, Shields (1989) arrived at a total of 17 280 described taxa but, again, there were some approximations in his listing. He claimed that a realistic estimate can be made only for the swallowtails (Papilionidae). The numbers in other large groups are probably over-estimated because many of the names may prove to be synonyms. In the past, many butterflies from unusual or outlying localities, or showing slight differences in colour, pattern or size, have received either specific or subspecific (trinomial) names, and it is commonly not at all clear whether some of these represent valid biological entities or merely trivial variants: their status cannot be clarified without biological information or longer series of specimens, and so remains debatable.

This ramification of the philatelic approach to butterflies has occasionally been carried to highly confusing extremes, as the fashion for collecting and naming 'varieties' or 'aberrations' of common species earlier this century sometimes resulted in tongue-twisting names which could apply only to a particular individual. These can be disregarded except as descriptive terms, and in evidencing a high degree of variability in the species concerned. One example of this trend is the plethora of descriptive names applied to forms of the British Chalkhill Blue butterfly (*Lysandra coridon*) by Bright and Leeds (1938). Such epithets as '*pallidula-infralavendula-fowleri*' or '*syngrapha-inframarginata-antiirregularia*' applied to varietal forms of a butterfly only some 30 millimetres in wingspan nowadays appear rather grotesque; yet they are indeed descriptive, and it must be remembered that many such names were not intended to have any real taxonomic validity. Such names, if they are shown to designate individual variants, are considered 'infra' categories, and are not covered by the Rules of Zoological Nomenclature.

But some varietal names do represent constant morphs or biologically consistent forms of particular butterflies. Some, for example, show distinctive seasonal forms so that (for temperate region butterflies) spring and summer forms and (tropical species) wet season and dry season forms may differ greatly in appearance on a regular sequence of generations appearing at different times of the year, sometimes to the extent that they have been named as separate species. Such 'seasonal polyphenism' (Shapiro, 1976) is an annually repeating pattern influenced by environmental factors. A range of studies on Hesperiidae, Pieridae, Nymphalidae and Lycaenidae have shown that both temperature and day-length during larval development may influence the final appearance of some temperate species. In the case of many Pieridae, for instance, the insects are darker (with more melanin pigment) during the cool season. The striking differences between some wet season and dry season tropical Nymphalidae : Satyrinae ('browns') is linked with substantial differences in behaviour (Brakefield and Larsen, 1984). Dry-season forms are often very cryptic and may aestivate, whereas wet-season butterflies are generally much more active and more strikingly marked. Brakefield and Larsen suggested that seasonal variations in ecological conditions may favour hiding and inactivity in the dry season, and a different kind of defence ('warning colouration') and higher reproductive activity in the wet season.

Warning colouration, put simply, is associated with advertising distastefulness to predators such as birds, which can learn to associate a particular pattern with a noxious or unpalatable prey. It often involves very bright colours (such as red, black, yellow, white) in contrasting bands. Large 'eye spots' on butterfly wings may also have the function of deterring or distracting predator attack. Some tropical butterflies are members of complexes of 'mimics and models' whereby palatable species closely resemble toxic or distasteful butterflies and so are protected if a predator learns to recognize the distinctive colours of the latter and avoids other insects which resemble this. Mimicry is a vast subject of considerable evolutionary interest and is mentioned here solely to indicate another aspect of butterfly variation (see reviews by Wickler, 1968; Turner, 1984). Particular rare and genetically stable morphs of a species may occasionally evince conservation interest.

Males and females of the same species can differ substantially in appearance. In other species the two sexes are very similar. Colouration in butterflies is associated with many aspects of a species' biology, and sexual dimorphism, as well as perhaps facilitating mate recognition, can reflect sexual differences in behaviour such as thermoregulation (Clench, 1966). Many butterflies which are characteristic of cooler climates, including high latitudes and high altitudes, tend to be dark (for widely distributed species, individuals in such populations commonly are both darker and smaller than others), thus enhancing their ability to raise their body temperature rapidly by absorbing the sun's radiant energy. Such strategies may enable butterflies to become active earlier in the day and stay active later, thus increasing their overall period of activity (Kingsolver, 1985a,b). Differences in the pigmentation of the two sexes virtually always appear to indicate differences in habits between them, a situation which in some ways parallels the seasonal variations just outlined. In hot open areas, such as some limestone areas of western Europe, the main problem may be overheating and lack of shade — in these environments some butterflies overcome this by being very pale, so that they reflect (rather than absorb) solar energy. Such habitat-related or climate-dependent variables are sometimes very difficult to interpret, and can together produce complex patterns of geographical variation in a given butterfly species. Many of these forms are likely to have been named. Taxonomic complexes such as the more than 650 named forms of the Apollo butterfly (*Parnassius apollo*) in Europe may indicate a substantial degree of real biological variation and incipient species formation.

Definition and diagnosis of such categories is fraught with difficulty, and uncertainty over the status of many of these means that some populations carrying a distinctive name may be real biological entities, or species, and others are not: the process of clarifying the status of these inevitably leads to variations in estimates of numbers of species. Shields (1989) cited the progress which has been made with just one group of Nymphalidae, the Ithomiinae, in which the recognition of many variable ('polytypic') species rather than more numerous ones which are constant in appearance has reduced the estimated number of species in the subfamily from more than 800 in 1909 to slightly more than 300 during the last few years.

Properly, a butterfly can have two or three words in its scientific name. These form a hierarchy. The first, always commencing with a capital letter, is the genus (or generic) name. The 'genus' comprises a group of closely related 'species', and each is denoted by a separate second name which always begins with a lower-case letter. A species (often loosely used to mean a 'kind' of butterfly) is defined as a population of individuals which can interbreed with each other but not with members of other species. With many butterflies, as with other insect groups, this practical definition is difficult to apply because the actual tests for reproductive compatability or isolation between members of different populations have not been made. Dead butterflies on pins have formed the basis for most designations of species. Geographical isolation is often taken to infer reproductive isolation, as members of widely separated populations may never meet. Many species are divisible into separate populations of this sort which differ more or less consistently in small features of markings, size or other parameters, but which specialists do not consider to be completely distinct species — although they may be on the path to becoming so. These are commonly designated by a third name — the subspecific name, also commencing with a small letter.

Related genera are grouped together in families, and related families in superfamilies. The following example shows the systematic position of an Australian butterfly of conservation interest, the Eltham Copper (see pp. 182–4):

Superfamily Papilionoidea
Family Lycaenidae
Genus *Paralucia* (the genus contains three species, all restricted to Australia)
Species *pyrodiscus* (Doubleday) (which contains two named subspecies)
Subspecies *lucida* Crosby (a name which differentiates it from the other subspecies, *pyrodiscus*). The name at the end is that of the person who first described the butterfly and gave it the name. If the author's name is in brackets, this means that the species has been transferred to another genus since it was described.

As implied above, the biological status of many butterfly populations which have been named as subspecies is not at all clear, and much of the confusion arises from our often naive attempts to pigeon-hole, or impose static nomenclature on complex and dynamically evolving groups of animals. Many such names at first appear to be applied sensibly, but further investigations may reveal that they are applied to parts of a sequence of continuous variation (a cline) in a species. The names may still be useful as labels but, as Bowden (1985) commented, it is not always wise to set 'the simplicity of an artificial classification above the biological complexity of evolution in time'. Classification consists of grouping similar individuals and also 'ranking' these into a hierarchy. The boundaries of groups and the precise level of ranking may both pose problems.

The classification of butterflies at the family level here is based on that of Ackery (1984), although many books on butterflies differ from this. As examples, some subfamilies of Nymphalidae (especially the Nymphalinae, Acraeinae, Danainae and Satyrinae) have often been accorded full family status (so that the ending '. . . inae' becomes '. . . idae') and the Riodininae are commonly treated as a distinct family (Riodinidae) rather than as a segregate of the Lycaenidae. There are other anomalies, but the taxa listed as of sub-family or family status in Figure 2.1 are generally each recognized as distinctive major entities. Differences between various accounts may seem confusing: essentially, there is still some disagreement about how some butterflies should

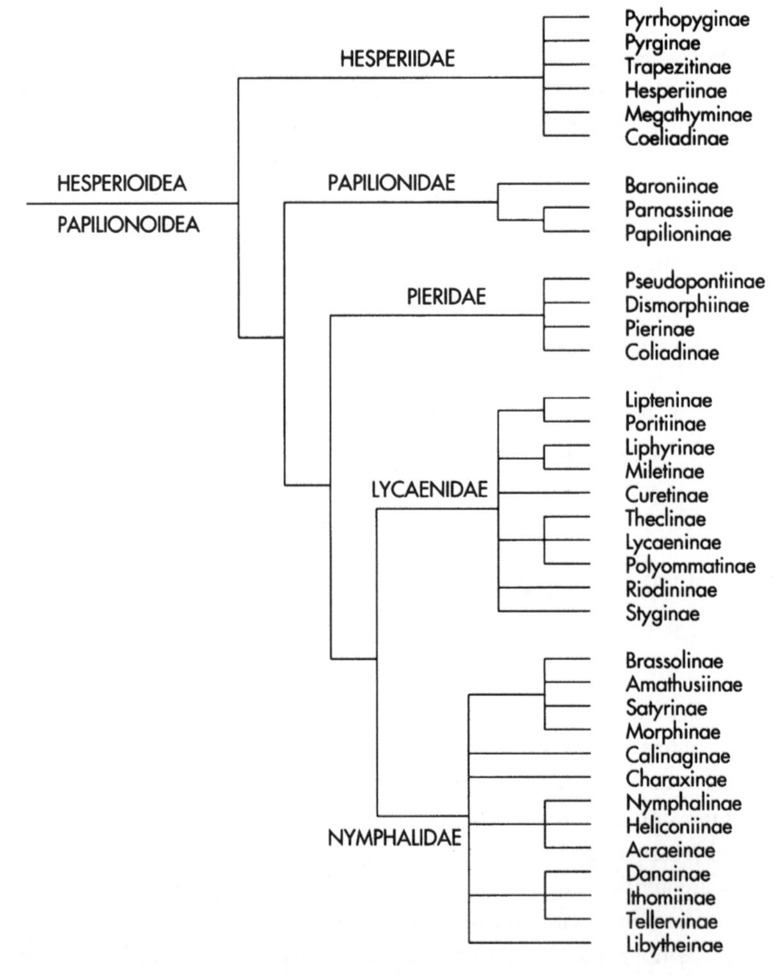

Figure 2.1 The classification of butterflies (after Ackery, 1984).

be classified at the family and subfamily level, but much of this need not concern us in practical terms. Each of these is clearly recognizable as a major entity comprising genera and species which are more closely related to each other than to those of any other butterfly group. The Riodininae, for example, resemble the other Lycaenidae in many structural and biological features yet is sufficiently discrete to form a separate grouping from these; whether or not it is then designated formally as a subfamily or a discrete family is largely immaterial as long as it is recognized as a group.

Some groups are very small, and some of these are recognized as rather archaic forms. Three, Baroniinae, Styginae (sometimes included in Riodininae) and Pseudopontiinae, contain only single species. At the other extreme, Nymphalinae, Satyrinae, Theclinae and Hesperiinae each contain more than 2000 named 'species' which may be valid, and Polyommatinae, Riodininae and Pyrginae each have more than 1000. Three families, Hesperiidae, Nymphalidae and Lycaenidae, together include more than 80 per cent of all butterfly species. Lycaenidae is often claimed to be the largest family, estimated to contain 30–40 per cent of all butterflies.

With practice, most butterflies can be recognized as belonging to these major groups with a high degree of reliability. The following brief synopsis of the five major families will help in this:

Hesperiidae The 'skippers', 'awls', 'darters', and 'flats'. These generally small and sturdy butterflies have antennae which are thickened slightly before the apex rather than having an apical club, and the tip is commonly reflexed or slightly hooked. The veins in the apical part of the wings are not branched, and all six legs are functional. Six subfamilies are recognized, and they total some 3500 described species. Hesperiidae is a very discrete group of butterflies and occurs in many parts of the world. It comprises the super-family Hesperioidea, and the remaining families constitute the Papilionoidea.

Papilionidae The 'swallowtails', 'birdwings', 'apollos' and related forms are the largest and most spectacular of all butterflies. They are predominantly tropical, but some species occur in cool temperate regions. The hindwings often have long 'tails' and all three pairs of legs are functional, with simple claws. About 570 species are divided among three subfamilies.

Pieridae The 'whites', 'sulphurs', 'orangetips', and 'jezabels' also have six functional legs, but the claws are bifid, or forked. Most species are medium-sized, with few larger than about 80 millimetres in wingspan. The 1000–1200 species occur in many parts of the world and are placed in four subfamilies.

Nymphalidae The 'browns', 'owls', 'morphos', 'admirals', 'fritillaries', 'milkweeds', 'beaks' and others are taxonomically very diverse, so that the Nymphalidae includes up to 13 subfamilies. The males have very reduced and clawless forelegs and their dense covering of scales has resulted in the common name of 'brush-footed butterflies'. The females' forelegs are also somewhat modified, but still have claws. Many of these butterflies are large and robust in appearance. More than 6000 species makes the Nymphalidae one of the largest groups of butterflies, and some of the subfamilies are geographically restricted.

Lycaenidae The 'blues', 'coppers', 'hairstreaks', and 'metalmarks' also have the male forelegs reduced but, except in Riodininae, these still retain a claw. The female forelegs are normal, and most members of this large group (up to about 9000 species) are rather small butterflies. Few are larger than about 50 millimetres in wingspan, and the family contains the smallest of all butterflies. The exact species that holds this distinction is not wholly settled, but *Micropsyche ariana* from Afghanistan is only about 7 millimetres in wingspan. This widespread family is divisible into as many as 10 subfamilies.

DISTRIBUTION

As I implied earlier, particular groups of butterflies are commonly geographically restricted, so that each major region of the world has a characteristic fauna. Indeed, very few butterflies occur naturally in all parts of the world, and most have a characteristic distribution pattern. As examples, trapezitine skippers are predominantly from Australia and New Guinea, Megathyminae are restricted to the southern USA and Central America, the birdwings to the South-East Asia – New Guinea area, Dismorphiinae and Riodininae are predominantly neotropical, Acraeinae are most diverse in Africa, Heliconiinae in South America and Danainae in the Oriental Region. Localized distributions or levels of endemism become more pronounced at lower taxonomic levels, so that a high proportion of Australian Lycaenidae : Theclinae (6/12 genera, 25/35 species) do not occur elsewhere (Common & Waterhouse, 1981) and many of the butterflies of southern Australia are restricted to that region. Such situations often reflect evolutionary isolation, where species can diversify without influence from outside populations. More 'open' environments, where ready exchange can occur with the butterflies of other regions, may not foster the development of endemism (species restricted to the area) to the same extent. For instance, although the UK has some subspecies (or incipient species) which do not occur elsewhere, virtually all the recognized species are shared with other parts of Europe (Dennis, 1977). On this larger scale, though, about a third of Europe's 380 or so butterfly species do not live in any other part of the world. Examples could be multiplied *ad nauseam*, but one of the widespread principles of species generation is that, over time, isolation promotes differentiation. This pattern means that, for many groups of butterflies, we can identify places of particular importance for their evolution and conservation — both on a broad scale and (progressively as better documentation occurs) on a finer, more local level. We can thus gradually highlight areas which support especially high levels of diversity and those with high endemism, and which it may be especially important to conserve to retain a good representation of the world's butterfly fauna. These have sometimes been designated 'critical faunas' (see Ackery and Vane-Wright, 1984, on milkweed butterflies, and p. 205).

It is hard to speculate on the long-term evolution of the butterflies as a group, not least because their fossil record is very poor. The Lepidoptera seemingly arose during the Cretaceous period somewhat more than a hundred million years ago, in parallel with the evolutionary expansion of the flowering plants, angiosperms. Butterfly fossils are extremely unusual, probably because they are of rather delicate insects with large wing areas. As just one example, of the 15 000 insect fossils from Colorado examined by the American lepidopterist S. H. Scudder (1889a), only eight were butterflies (Shields, 1976). Many of the 50 or so known fossils are from the Oligocene period, and it appears that all the major butterfly families had evolved by then. A few butterfly fossils are assigned to modern genera and some even to present-day species. Most speciation in butterflies occurred up to the early Pleistocene period (about 1–10 million years ago), and most later development may have been of subspecies and intraspecific forms (Brown, 1987).

BIOLOGY

Not surprisingly, butterflies show a great diversity of biological features, as variations upon a rather simple basic life cycle. They have the kind of life cycle known as a 'complete metamorphosis' (Figure 2.2, p. 20). After mating, the female lays eggs. These hatch into larvae (caterpillars) which feed voraciously and grow via a series of moults. There are commonly three to six larval stages, or instars. Caterpillars eventually transform into a resting stage, the pupa or chrysalis, where the body is broken down and reorganized into the adult form. The adult butterfly emerges from the pupa to complete this widespread but wondrous cycle shared with most of the other more advanced insect groups. Variations, often limited to a particular group or species, include such things as adult behaviour (distance of dispersal, times of flight, feeding habits, mate-seeking and courtship), oviposition (whether eggs are laid singly or in groups, whether the oviposition site is selected actively or eggs are deposited more casually, length of oviposition period), larval feeding (whether only particular kinds of food are eaten or whether they are relatively generalist feeders, feeding by day or night or both, length of developmental period) and other behaviour (defence against predators, solitary or gregarious), pupation (position of pupa — suspended from the food plant or lying on the ground [*Baronia* even tunnels in the ground to pupate], cryptic or exposed, close to larval food or further away) and adult emergence (time of day or year, time elapsing to reproductive maturity, length of life, whether the two sexes emerge together or one before the other), and so on. The combinations are innumerable, yet characteristic of given

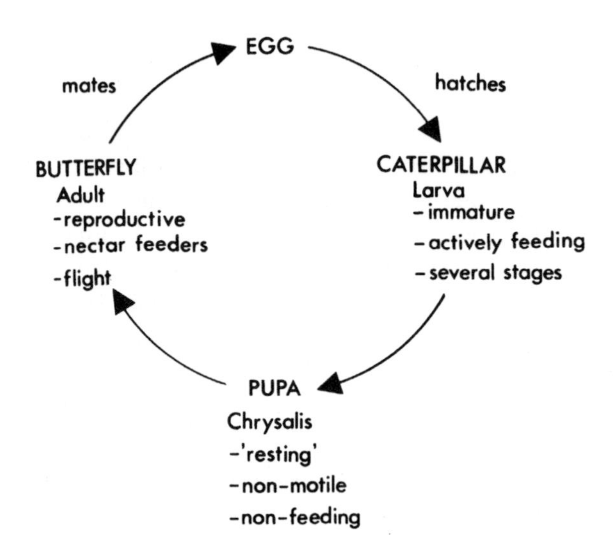

Figure 2.2 The life cycle of a butterfly, illustrating the stages of a complete metamorphosis.

butterflies, so that even closely related species (or subspecies) may differ in significant aspects of their biology. The life history strategy of each species tends to represent an optimum use of resources (which may be available for only part of a year and/or scattered in space) in climates which are seasonally variable and which may restrict periods of feeding or activity to more clement parts of each year. Alpine species may pupate on the ground to overwinter while their relatives in other areas may pupate on plants. The particular features of each species, then, show how it fits into its environment and demonstrate its resource needs and trends towards specialization. Knowledge of many of these, as we shall see, is needed in order to plan for conservation of a species and to ensure its continuing wellbeing as environments and habitats change. The remainder of this chapter indicates some of these variable facets of butterfly biology.

Most butterflies in temperate regions have well-defined 'seasonal' strategies, although the number of generations a year varies between species, or in different parts of the range of widespread ones. Many species are univoltine (having one generation each year), others are bivoltine (with two generations a year) or more. Some tropical species are relatively aseasonal and may breed continuously throughout the year because their development is not interrupted by periods (such as cold winter) when foodstuffs are unavailable. In temperate regions, most butterfly species overwinter as eggs or pupae in a state of arrested development (diapause or other forms of dormancy), but a few hibernate as adults. A similar resting phase may be used to overcome very hot times in some other parts of the world, although such regular 'aestivation' seems to be quite rare among butterflies.

For many species, even the number of generations usual in each year is by no means clear. One confusing factor is the length of life of adult butterflies. Many live for only a few weeks, or even less, and so have very well-defined flight seasons giving minimal confusion over interpreting this phase of their life cycle. In contrast, some others may live for many months so that their persistence, sometimes coupled with periods of inactivity or inconspicuousness, may give the illusion of more than one generation each year even though they may be truly univoltine. Local variation in life cycles can reflect climatic differences, often linked with the availability of food plants.

In the UK the Common Blue (*Polyommatus icarus*) has two distinct generations a year in southern England but is univoltine in northern England and Scotland. The boundary between these two patterns, where there may be some form of intermediate pattern, has apparently not been studied (Pollard and Yates, 1993). The single flight season in the north commences before the second flight in the south, and may be much longer than either of the bivoltine flights (Figure 2.3).

In coastal California, several univoltine species have very long flight periods, reflecting mild winters and cool summers with a staggered development of larval food plants (Langston, 1974). Their inland counterparts, often the same species, in the Coast Range have much shorter flight seasons, correlating with increased elevations and more rapid and uniform development over only a relatively short period. Many 'text book comments' on the times of appearance of particular butterflies are very general and can be modified considerably for specific localities within a broad distribution range. A practical inference for conservation assessment is that, to paraphrase Charles Kingsley's comment on waterbabies, a butterfly has to 'be seen not to be existing' before it is claimed to be absent or to have disappeared, and visits at even slightly out-of-phase times of the year may give ambiguous results. Often this provision is not feasible from brief surveys, especially in unusual climatic regimes and for species whose biology is poorly known, but simply determining the presence or absence of a rare species can be difficult (p. 89).

The swallowtail *Papilio alexanor* in Palestine has caterpillars which feed only on the flowers of umbellifers which are adapted to arid regions and bloom only in years of sufficient rainfall. In common with some other arid zone butterflies, this species can undergo a period of arrested development in the non-feeding pupa stage which can last at least two or three years, so that it would not be evident at all in years when the food plants are not available (Nakamura and Ae, 1977). Another, related, context is one where voltinism depends on the relative degree of food plant specialization. *Papilio zelicaon* is univoltine in areas where

Figure 2.3 Seasonal flight patterns of the Common Blue, *Polyommatus icarus*, in the UK, to illustrate geographical variation in development (after Pollard and Yates, 1993). Abundance is indicated by histograms over the April (A) to September (S) period, with small dots denoting absence: there is a clear transition from northern univoltine to a southern bivoltine pattern.

it feeds on native Umbelliferae in California, but other populations utilizing introduced food plants are multivoltine (Sims, 1980). Food plants here also influence the diapause regime, which is related to food quality and availability. The resources needed by a given butterfly species are sometimes very precise, so that environmental change can disrupt its wellbeing in many very subtle ways.

The adult is the mobile dispersive stage in the life cycle in relation to the caterpillar, which can normally only walk short distances and may move around very little during development. In general, adult feeding habits tend to be less specific or restricted than those of the larvae, although some patterns do exist. As Gilbert and Singer (1975) noted, butterflies are known to feed predominantly on nectar from flowers, but also on or from honeydews, froghopper secretions, rotting fruit, sweat, urine, dry or wet dung and carrion — rotting meat, corpses or fish — baits often employed by collectors to obtain otherwise elusive butterfly species. On dry surfaces, butterflies may egest saliva onto the potential food to soften it and facilitate ingestion, as they can feed only on liquids; in this way they may wash out sugars from rather inauspicious-looking substances. There has been very little experimental work on food selection, but Gilbert and Singer pointed out some taxonomically-based generalizations which appear to be valid. Danainae feed from flowers or some kinds of dead foliage, Ithomiinae on nectar and bird droppings, and Charaxinae on dung and rotting plants. Some other groups of Nymphalidae are not flower visitors. Diurnal flower-feeders in the tropics tend to be small or hard to catch, distasteful and warningly coloured, or mimics of distasteful species. In contrast, dung, fruit and sap feeders are commonly both edible and cryptic in appearance. Most, though, can easily move around to seek food.

Because of the relative immobility of caterpillars, it is very important that the female butterfly lays her eggs on or close to suitable food, and because many kinds of caterpillars characteristically feed on only one plant (or a very limited range of plants), the selection of oviposition site is a crucial process. Especially in the case of extreme feeding specialists, the female butterfly may condemn her offspring to starvation and death unless she lays her eggs in 'the right place'. The topic of selection of oviposition sites has recently been reviewed by Chew and Robbins (1984), who point out that many female butterflies do indeed make such mistakes as part of their normal behaviour! What constitutes a suitable food plant for caterpillars and what constitutes a suitable egg-laying site may not necessarily be the same. However, some such apparent errors may really not be so — one case cited from time to time is of the Silverwashed Fritillary (*Argynnis paphia*) in the UK, where the females may lay eggs on tree trunks rather than on the larval food plant, violets. The caterpillars in fact rest during winter in bark crevices, and then drop to the ground in spring to seek food (Singer, 1984). Any apparent anomalies in behaviour need careful attention and interpretation.

Larval feeding is determined predominantly by plant chemistry and gross features of the plant surface, such as degree of hairiness or toughness — features regarded respectively as 'chemical defences' and 'physical defences' of the plant against insect attack. Adult butterflies may be attracted to lay by the presence of various plant chemicals, but 'preferences' may vary with the age or nutritional state of the female, as well as season or geographical location; such variations emphasize the need for caution against overgeneralizations. Females may, for example, be attracted to particular shades of green (*Pieris brassicae*: Ilse, 1937), to particular leaf shapes (*Heliconius*: Gilbert, 1975) and other visual cues, and be able to recognize these. Such responses may be influenced by the degree of contrast between the food plant and surrounding vegetation, or the relative abundance of food plants and other plants, and it is likely that female butterflies learn to associate leaf shape with other plant characteristics. They could then utilize easily-recognized visual cues to increase their searching efficiency in natural plant communities by targeting particular suitable larval food plants. Many butterflies (in the Papilionidae, Pieridae and Nymphalidae) are known to avoid laying on plants which already bear a complement of eggs from earlier visitors. This is seemingly one mechanism for avoiding possible competition for food among the offspring, but it may also ensure that most individual food plants present are exploited to some extent.

Discrimination between individual plants of the same species is a complex and little-understood theme, but is one of considerable importance in conservation. Some of the cases discussed in Chapter 7, for example, show situations where particular ages, stages or growth forms of plants are acceptable for oviposition whereas other categories of the same species in the same locality are either ignored or do not sustain caterpillar development. Likewise, the quality of similar-looking plants can vary with soil factors or in relation to qualities of neighbouring individuals and species. As Courtney (1984) emphasized, another aspect of oviposition site selection is the ease with which host plants are available in relation to the time spent looking for them. It follows from conventional optimum foraging theory (whereby an organism should maximize its returns for a given effort in seeking resources) that a female with limited time to lay her eggs or search for a suitable place should include as many food plant kinds as possible. In contrast, if time is not restricted, a female (or species) may be able to afford to specialize to a greater extent and search for particular plant species or other categories. A complex sequence of behaviour, utilizing a range of different stimuli, may be involved in a female butterfly finding the right oviposition site (Figure 2.4, Courtney, 1986).

Distribution of eggs on food plants also varies, and the patterns may reflect both movements of laying females and distribution of plants. Peripheral and isolated plants may be particularly heavily exploited, possibly because these are more easily seen or are more accessible than the centre of denser plant stands. The deterrent effect of existing eggs may result in more regular egg deposition across all available plants. A number of intriguing experiments have shown that visual cues are involved in this, and some food plants of *Heliconius* apparently protect themselves from oviposition (and hence from the ensuing caterpillar attack) by developing small swollen yellow pustules which superficially resemble eggs. These egg-mimics are sufficient to deter egg-laying by searching *Heliconius* females.

Most butterflies lay their eggs singly, but some oviposit in batches or clusters, and *Hamadryas* and *Araschnia* (Nymphalidae) lay eggs joined in 'strings' standing up from the food plant. Clusters can sometimes contain eggs from more than one female. Egg-clustering has been recorded in species of most butterfly

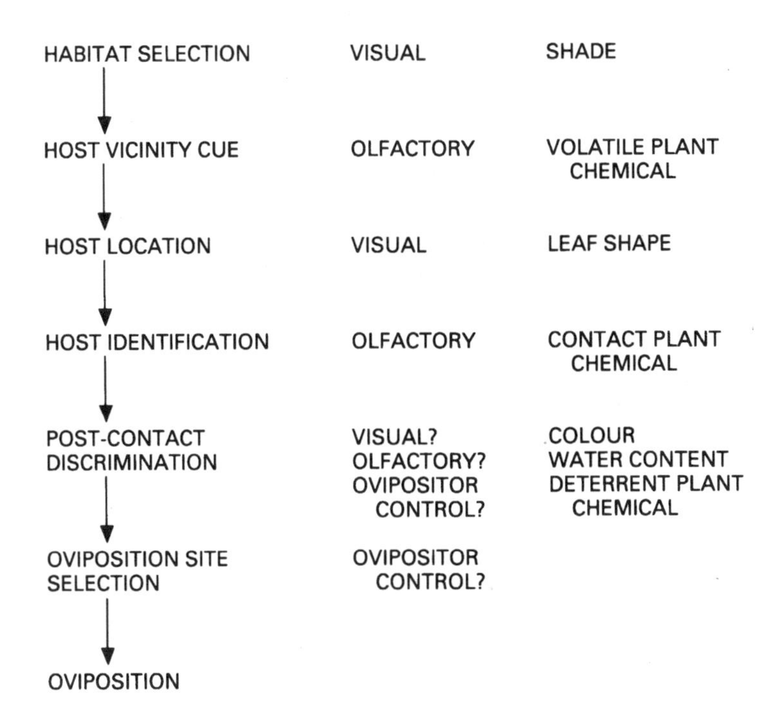

Figure 2.4 Stimuli involved in selection of oviposition sites by butterflies. The stages of discrimination are shown on the left, the senses utilized in the centre, and examples of cues needed on the right (Courtney, 1986).

families. Often, but by no means always, such eggs are distasteful to predators, and in both egg clusters and gregarious caterpillars their mutual proximity may help in defence against predators. In most cases the female butterfly shows no further interest in her eggs once they are laid, but those of *Hypolimnas antilope* (Nymphalidae) stand guard over their egg batches until they hatch — a highly unusual example of parental care in the butterflies.

Most caterpillars are herbivores, voracious chewing feeders on plant material of various kinds. Many species, as we have implied, are specialists and can feed only on one kind, or a few related kinds, of plant. Others are relative generalists with a broad food spectrum, but most are to some extent restricted in the range of food plants which they can exploit.

Other caterpillars, including those of many Lycaenidae, have departed from this plant-feeding habit to varying extents to associate with ants and, sometimes, to feed entirely on ant larvae and pupae. The biological intricacies of many myrmecophilous Lycaenidae are highly specific. Many species are highly localized, with their restricted distributions reflecting the specialized needs of their host ant as well as those of the food plants. They are commonly predominant amongst butterflies needing conservation management in many parts of the world (New, 1993), not least because their populations can be very patchy and restricted in incidence. For some species, the host ant itself is rare and, in a wider perception of conservation, would also be a target species for management in its own right, rather than simply as a critical resource for a butterfly. One reason for the great diversity of Lycaenidae may be their expansion to this specialized but abundant resource, and consequent involvement in suites of ecological interactions not open to other butterflies. Female lycaenids often use ants as oviposition cues, in addition to features of the host plants on which their caterpillars may partially depend. Some perhaps detect the presence of suitable ants by testing foliage with their antennae to detect the chemical traces or scents of ants before they lay their eggs. Some, indeed, go further and lay among the ants with the aid of appeasement chemicals which reduce the likelihood of attack while doing so. Still others use ant-attended insects, predominantly aphids or other sap-sucking bugs which produce honeydew used by the ants, as oviposition cues. Caterpillars of some Miletinae feed on these bugs themselves, as another example of expansion of ecological range in the Lycaenidae. For some lycaenids it may be more important to maintain a particular specific ant association than one with a particular food plant (Pierce, 1984) so that experimentation with oviposition could lead to expanding the host plant spectrum if the caterpillars remain partly dependent on plant food.

It is very difficult to generalize about the food plants used by butterfly caterpillars, although there are numerous published records of these. Since Ehrlich and Raven (1965) introduced the idea of co-evolution between butterflies and host plants mediated largely through features of plant chemistry, the literature on this topic has burgeoned. Most species are restricted to angiosperms, and very few are known to feed on ferns. One species of *Euptychia* (Satyrinae) feeds on lycopsids in Panama (Singer *et al.*, 1971), and some Lycaenidae (Lipteninae) feed on algae, fungi or lichens (Cottrell, 1984). Some riodinines feed on epiphylls on a taxonomically diverse range of host plants (de Vries, 1988a), and de Vries suggested that this habit might be more widespread than known in this group. Otherwise, though, the 'lower plants' are not known to be attacked.

Chemical feeding cues, which stimulate a caterpillar to continue eating (or, conversely, repel it) once tasted, may vary in incidence with the age of the plant (so that caterpillars may be restricted to feeding on either young foliage or older leaves). Or there may even be diurnal cycles of chemical incidence, influenced by sunlight, so that some caterpillars (including various Nymphalidae) may have to feed at night. Many plant chemicals are restricted to particular species, genera or families of plants and play a major role in feeding specificity.

Many plant–butterfly associations may be of considerable antiquity. Several fossil butterflies (such as the Eocene *Prepapilio colorado*) are clearly allied to primitive and taxonomically isolated Recent taxa which are specialized feeders — in this case to the Mexican *Baronia brevicornis* (the only member of the Papilionidae : Baroniinae) whose caterpillars feed on *Acacia*. Zeuner (1962) regarded the *Aristolochia*-feeding *Troides* swallowtails as the most primitive living butterflies. Nevertheless, many kinds of butterflies have a very limited spectrum of plants on which their caterpillars can feed, and many collectors rapidly learn to recognize these when searching for larvae of particular kinds.

Some broad generalizations on caterpillar food plants can be given, albeit rather tentatively. As a family, the Hesperiidae feed on many different plants, though most species are relatively specific. Some species of *Coeliades* in Africa are polyphagous, so that one species has been recorded from five families of plants (Pinhey, 1965). In contrast, some Australian Trapezitinae, predominantly grass or sedge feeders, seem to be restricted to particular species of *Gahnia*. Skipper caterpillars commonly roll up a leaf or join several leaves together with silk so that they live inside a tube or under a web. The Nymphalidae : Morphinae and Satyrinae are primarily associated with monocotyledons (grasses, sedges, bamboos) but, whereas no Satyrinae apparently feed on dicotyledons, most species of *Morpho* (the most diverse genus of Morphinae) occur on these plants. Most other butterfly groups are limited to dicotyledons. Only fragmentary information is available for Hedylidae: one species feeds on Sterculiaceae (Kendall, 1976; Scoble, 1986) but more data are available for other families. As noted, Troidine swallowtails (Papilionidae) feed predominantly on Aristolochiaceae, as do most genera of Parnassiinae, and many other swallowtail caterpillars eat foliage of Rutaceae or a closely related family. In the Pieridae, Dismorphiinae feed predominantly on Leguminosae, as do many Coliadinae; Pierinae are more varied, though many genera feed on Cruciferae or Capparidaceae. *Delias*, a large Indo–Australian genus, feeds mainly on mistletoes and the related Santalaceae. Nymphalidae : Ithomiinae eat Solanaceae or Aristolochiaceae, both of which are rich in alkaloids. The related Danainae specialize on Asclepiadaceae ('milkweeds') and Apocynaceae, which seem to be closely related plant groups.

Many Lycaenidae are also specialist feeders, and the literature on this family is encyclopedic. Biological groupings in the family reflect the varying degrees to which lycaenids have become aphytophagous and dependent on animal food (Henning, 1983; Cottrell, 1984). Most genera in the largest subfamilies, Polyommatinae and Theclinae, appear to contain species which are opportunistically carnivorous. *Maculinea* (p. 171) and *Lepidochrysops* caterpillars feed solely on plants during their early instars but later feed only on ant larvae or pupae. Liphyrinae and Miletinae have evidently abandoned plant feeding altogether, and isolated taxa in some other groups are also entirely myrmecophagous.

The distribution of larval feeding patterns within any family is often hard to interpret, but in the Papilionidae much higher relative proportions of species in temperate regions are generalist feeders and most species close to the equator are specialists (Scriber, 1973). The distribution and availability of food resources, be they particular kinds of larval food plant, particular ant or bug taxa for lycaenids, or flowering plants for the nectar-feeding adults, is an important correlate of butterfly distribution, and, of course, climate may play a major part in determining these. As a generality, butterflies are most abundant and diverse in the floristically rich lowland tropics, and some major groupings do not occur, or are very scarce, elsewhere. They become progressively less diverse at higher latitudes and higher altitudes, and many butterflies of those more extreme environments are not found elsewhere. Within these gross regions, particular butterflies are often restricted to certain kinds of vegetation associations or microhabitats so that, as examples, species can be allocated as 'forest or woodland species', 'forest edge species', 'open glades in woodland species', 'heathland species', 'grassland species', 'scrubland or 'hedgerow species', 'marshland species', 'sand dune species', and so on. For better-known faunas, most species can be allocated to habitat categories in this way, and knowledge of such associations is valuable in conservation assessment. Each major vegetated terrestrial biome can support a characteristic complement of butterfly taxa. Within each one, particular species, especially those which are ecological specialists (Table 2.1), may be very localized. In Australia, for instance, a number of endemic Satyrinae in the south-east are regarded as alpine or subalpine, and are restricted to particular bands of altitude and open heaths or button-grass plains in these limited highland areas. A definitive recent account of Swiss butterflies (Ligue Suisse, 1987) gives such details for virtually all the 175 native species. In mountainous country in many parts of the world, discrete colonies of butterflies may occur on isolated highlands, and many such populations have been accorded subspecific status, because each has diverged in appearance from the original parent species to become a consistently distinct form, and individuals may not be able to move between colonies to maintain a continuous gene pool.

Table 2.1 Predicted patterns of niche structure and adaptive strategies for butterflies in different types of habitats in the tropics (Young, 1980).

Habitat	*Spatial patchiness of host plants*	*Relative patch size*	*Larval feeding pattern*	*Adult feeding pattern*
Forest canopy	very high	small	polyphagous	generalist
Forest understorey	high	small	polyphagous to monophagous	generalist
Forest understorey	low (palmaceous)	large	monophagous	generalist
Forest light gap*	high	small	polyphagous	generalist
	low	large	monophagous	generalist
Secondary succession	low	large	monophagous	specialist

* The forest light gap represents a highly transitional habitat between the highly patchy forest flora and the less patchy flora of secondary succession. Two cases are recognized here, although most light gaps very likely are mixtures of both.

DISPERSAL

Isolation, dispersal, and the functional relationships between these, are important parameters of the ecology of a butterfly species and are of considerable relevance to conservation, as we shall see later. Because we commonly see butterflies flying, we tend to think that adults of all species can disperse and colonize 'new' areas without difficulty, but this assumption is often not justified. Many lycaenids, for example, are remarkably sedentary: *Agriades pyrenaicus* in the Ukraine has not been observed further than 50 metres from its usual habitat (Pljushtch, 1989), and some species seem to be even more immobile than this, even in a habitat with no obvious 'barriers' to flight. At the other extreme, some butterflies are well-known migrants. The Monarch (*Danaus plexippus*) in North America expands from overwintering areas in California and Mexico (see p. 161) to colonize much of North America during the warmer part of the year. The same species is known as the Wanderer in Australia, and was first encountered there around Brisbane (Queensland) in 1870 (Miskin, 1871; Zalucki, 1986). It apparently arrived by 'island-hopping' across the Pacific Ocean (perhaps aided by human transport) and now occurs in New Zealand, New Guinea, Hawaii and other places, and is a very conspicuous element of the south-western Pacific butterfly fauna. *D. plexippus* is occasionally recorded in Britain or western continental Europe, apparently as a migrant across the Atlantic Ocean.

Many other butterflies exhibit a regular seasonal migration (compendia by Williams, 1930, 1958; Baker, 1968) during which they may cross substantial sea or land barriers to reach and exploit new habitats. Others may be dispersed inadvertently by winds, 'chance' or human agency; it is often difficult to separate these from regular migratory events other than by their rarity or sporadic nature. For one example we can consider the Krakatau Islands, between Java and Sumatra and about 40 kilometres from either of these source areas, which were sterilized by the famous catastrophic volcanic eruption of 1883. By 1986 they supported at least 55 kinds of butterflies, all of which must have colonized since 1883. Intermittent collections from about 1908 onwards have revealed that the number of species has increased as the vegetation has become more complex. Early collections included species which could not breed on the islands because no suitable larval food plant was present (New *et al.*, 1988). Opportunities to assess habitat colonization by butterflies on this scale are indeed rare, and it is presumed that most — even all — of the Krakatau butterflies arrived on the islands by natural dispersal from Java or Sumatra.

The fauna of Anak Krakatau demonstrates the rapidity with which butterfly assemblages can be formed. Anak is a young island which emerged from the sea only in 1930 in the middle of the Krakatau caldera, and is of particular interest in colonization studies as being one of the very few isolated tropical sites in which these processes can be assessed from their very earliest stages. By 1985, 23 species had been recorded. An additional 13 were present in 1989, and five more in 1990 raised the total to 41, some of which are not known elsewhere on the Krakataus (New and Thornton, 1992). Most were relatively generalist species, and none was a true forest butterfly. Geographical and (p. 51) habitat islands are a key theme in assessing population dynamics and the adequacy of reserves for butterflies.

A very different example is a large brassoline nymphalid, *Opsiphanes tamarindi*, which is native to central America (Mexico to Peru) and has recently been noticed on several occasions in Britain (Burton, 1986). This may be an accidental import, as larvae feed on banana foliage (among other things) and pupae could be imported in boxes of fruit. Instances of butterflies being found far from their natural range can be confidently attributed to artificial importations rather than any natural dispersal.

Much of Hawaii's butterfly fauna may have arrived there by 'hitching' on aircraft in recent years (Riotte and Uchida, 1976). Whereas *Opsiphanes* is unlikely to become established in Europe, other butterflies have done so in various parts of the world. *Pieris brassicae*, the Large White of Europe, is a pest of crucifers in Chile, for example (Gardiner, 1962), and *P. rapae* has the same status in North America and Australia. There are a few cases of deliberate introductions of butterflies to areas well outside their normal distribution: two species of Lycaenidae were taken to Hawaii early this century as potential biological control agents for lantana.

It may at times be very difficult to generalize even about the dispersal ability of a single species, because some taxa seem to consist of populations with different intrinsic dispersal capabilities. The British Swallowtail (*Papilio machaon*) is one such species; the American Edith's Checker-spot (*Euphydryas editha*) is another.

Clearly, localized populations of butterfly species which do not disperse widely may be vulnerable in many ways. If a butterfly disperses readily on a local scale, even if it is not migratory, damage caused to localized populations by some form of environmental change may be offset quickly by recolonization from nearby areas. In contrast, for sedentary species with discrete separated populations this may not be possible, and local extinctions may result. There is a close relationship between dispersal and vulnerability which manifests itself in several ways, not least through isolated populations often having a narrower range of adaptive ability to counter environmental changes because their gene pool is also isolated rather than being continually (or intermittently) 'refreshed' by individuals mating with invading conspecifics.

Intrinsic barriers to dispersal may be as simple as a few metres of open space or, conversely, intrusive physical features such as bushes or changes of gradient. Narrow bands of unsuitable terrain may, therefore, act as very effective barriers, but these are often difficult to assess. Behavioural trends may also limit dispersal.

In a population of *Euphydryas editha* studied by Ehrlich and his colleagues (Ehrlich, 1961), emigration (dispersal out of the population, in contrast to immigration) was found to be an inverse function of density of individuals, so that many individuals have a tendency to stay with other individuals rather than to move away from them.

These varying abilities or propensities for flight have been quite accurately assessed for many species by mark-release-recapture techniques (p. 95). Adult butterflies are captured and marked in some way — most commonly with spots of coloured paint or with felt-tip pens, more rarely by attaching numbered labels to the wings — and then released. The butterflies are then individually recognizable and if captured again, even some distance away, can be unambiguously identified. The recapture rate for long-distance migrants, such as the Monarch (in which clarification of its migratory routes involved a long-term labelling programme: Urquhart, 1960 and later works), is very low, so that it is necessary to mark large numbers of individuals to get worthwhile information. The level of dispersal activity can be used, in conjunction with the availability of food resources, to define several categories or classes of butterfly populations.

POPULATION STRUCTURE

Many species which disperse little have very specialized ecological requirements and may, for instance, be restricted to particular geological formations or soil types supporting characteristic vegetation. These commonly form geographically localized populations with little or no interchange between even discrete colonies separated by only a few hundred metres. Such populations are referred to as 'tight' or 'closed'. Greater levels of mobility between loosely defined segregates, means that many more dispersive species form 'loose' or 'open' populations which are inevitably less discrete. The former are commonly those at greater risk of extinction, since conservation of a circumscribed tight population or species may necessitate preservation and maintenance of the limited habitat. A closed (or tight) population may, though, extend over a considerable area, sometimes with the butterflies having a very low density. Thomas (1984) noted that some rather mobile British woodland butterflies (including the Purple Emperor, *Apatura iris*, and the Brown Hairstreak, *Thecla betulae*) breed at low densities over wide areas, but the boundaries of these areas are reasonably distinct and — despite the distances involved — adults congregate to mate at a particular point or points. Such behaviour maintains a closed population.

Pollard and Yates (1993) made the important distinction between 'matrix species' and 'island species' among the UK butterflies. In general, the 'matrix species' are relatively common and occur widely in the countryside matrix of fields, hedges, and copses, whereas the 'island species' are restricted to isolated habitats such as heaths, fens, larger woods or unimproved grasslands within the overall matrix. As in other parts of the world, it is predominantly the latter group which have been targeted for conservation as species, reflecting their restricted distributions and vulnerability to habitat change.

Most such target species are (1) ecological specialists, (2) highly restricted in range, at least within a given political or administrative region, and (3) perceived to be vulnerable or directly threatened by local changes. However, some species are difficult to allocate reliably to one or other of the main Pollard and Yates categories: a few of the British species, for example, occur as matrix taxa over much of their range but become more obviously restricted to 'islands' in places. A wide distribution of a butterfly, or other organism, may reflect one or more of the following (Dennis, 1992):

- tolerance of varying conditions,
- uniformity and ubiquity of required conditions, or
- wide dispersal capability.

Commonly, all three parameters appear to be involved, in contrast to the restrictions implied by highly localized distributions. Nevertheless, as Dennis (1992) noted, the parameters of a distribution can be measured and compared, as an avenue to understanding significant aspects of the species' biology and rationalizing conservation. Thus, distributions may differ in location, size, shape, orientation, various gradients (such as altitude, latitude, climatic or vegetational) and texture (density, contiguity or dispersion) (Dennis, 1992).

Relatively sedentary butterflies tend to be those which use host plants such as trees, which remain suitable for many generations; these have little need to move large distances, and tend to have closed population structures. Although they may fly actively, most of their movement is trivial and does not result in any change of habitat (Shreeve, 1992).

About 85% of British butterflies apparently form closed populations (Thomas, 1984) and, although such detailed information is not available for most other parts of the world, this figure can be taken to reflect the proportion of butterflies in some temperate regions which may be especially vulnerable to environmental change and so be directly relevant in conservation matters.

Small populations may be especially vulnerable to extinction by accident, such as localized environmental disturbance, and if such events are combined with low levels of dispersal and ability to colonize, the range of the species may contract. However, the concept of 'metapopulations', whereby a species exists as a 'population of populations' each occupying a small area of habitat but sustained collectively by new founding populations balancing extinctions of established ones (Levins, 1970; Harrison *et al.*, 1988), has become an important consideration in butterfly conservation in recent years. This type of population structure results in a 'shifting mosaic' distribution. Thus, *Euphydryas editha bayensis* (p. 165) has small local populations prone to relatively frequent extinctions caused by the effects of aberrant weather on the synchrony of the butterfly's life cycle and host plant. Harrison *et al.* (1988) showed that several habitat patches were colonized during their study, either for the first recorded time or after documented extinctions, with colonization reflecting the distance of source populations.

Thomas (1995) believed that most of the UK butterflies regarded as having closed local populations exhibited metapopulation structure, rather than being single isolated populations. This type of structure is characterized by:

a occasional movements between local populations (each a discrete demographic entity),
b colonizations and extinctions, and
c local populations occurring in groups rather than as single isolates.

He was able to demonstrate individual movements of butterflies between local populations, the extent depending on the species, distance between patches, and duration of observations. An example (the Silver-studded Blue, *Plebejus argus*) is given in Figure 2.5, from a more extensive study (Thomas and Harrison, 1992) in which nine metapopulations of this species were mapped at a seven-year interval (1983, 1990). The persistence of the species in the area

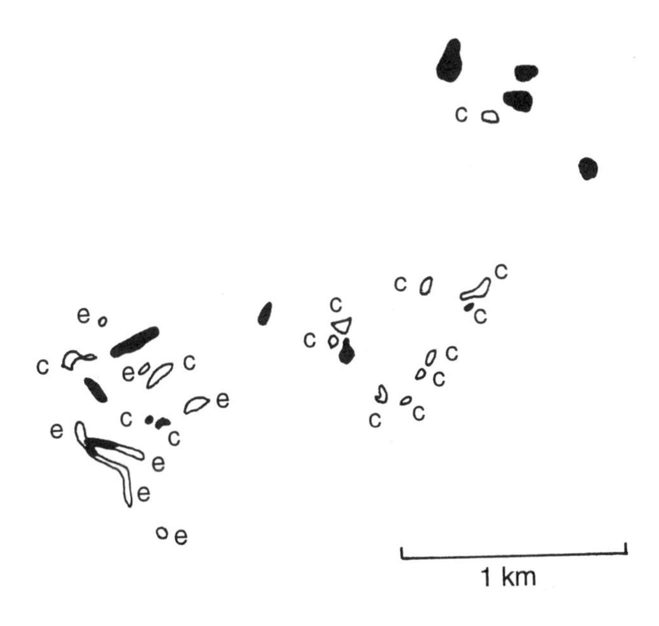

Figure 2.5 Example of a metapopulation structure reflected in spatial dynamics of the Silver-studded Blue, *Plebejus argus*, in a northern Wales locality. Distribution is shown for habitat patches (outlined) in 1983 and 1990: solid outlines = butterflies present on both occasions; empty outlines = not present on both occasions; e = presumed extinct (present in 1983 only); c = presumed colonization (present in 1990 only) (Thomas and Harrison, 1992).

depends on some suitable habitat (not necessarily the same patches persisting throughout) being continuously available within relatively large areas of biotope, and within a km or so of an existing population. Such structures reflect the transient nature of successional habitats. Some populations of *P. argus* in heathland act in this way, but turnover of habitat was slower in limestone grassland in which suitable habitats can be maintained for longer periods (Thomas and Harrison, 1992).

Within such a structure, populations will clearly vary in size, so that larger populations will be supported by larger habitat patches (Hanski and Gilpin, 1991; Hastings and Harrison, 1994), and such populations may be especially important in long-term persistence. These large populations may be relatively permanent and may serve as a source for colonization of nearby 'islands' of habitat, and much of the rolling sequence of colonizations and extinctions may be restricted to those smaller patches (Thomas, 1995). Appreciation of the dynamics of metapopulations becomes highly relevant in planning conservation in fragmented landscapes. For example, some poor-quality habitat patches near a source population may be populated through repeated immigrations rather than from a true resident breeding population over more than a few generations: the Marsh Fritillary (*Eurodryas aurinia*) exhibits such dynamics in the UK (Warren, 1994). As another example, most reserves are too small to support an entire metapopulation, but are clearly of value in harbouring source populations as refuges in surrounding unsuitable landscape.

However, as Thomas *et al.* (1992) indicated, even the largest populations (reflecting the largest habitat patches) may not be sufficiently large to obviate the likelihood of extinction for some butterflies, and the importance of 'stepping stones' or 'networks' of habitat patches sufficiently close to enable metapopulations to persist then becomes even more important. Thomas (1994) therefore recommended that management for such species should maintain existing habitat connectivity, or should restore connectivity by providing new habitat stepping stones, both of which will favour many other biota, in addition to butterflies. But the major change in attitude brought about by an appreciation of metapopulation structures is that 'single site' protection may not be adequate in itself for sustaining many butterfly species. Indeed, as Thomas (1995) noted, the presence of additional nearby populations of a rare butterfly is often used to justify the development of a site. A metapopulation structure suggests that every patch of habitat is potentially important, so that such sites may well need to be conserved to ensure the connectivity of the individual populations.

The need for 'mating points' by many butterflies is an intriguing aspect of their behaviour, and is very important to conservation planning. Many species exhibit the phenomenon known as 'hill-topping' (Shields, 1968), where males congregate at a high topographical point to seek mates. Often this is literally a hill top, but sometimes it is the tallest 'sentinel' tree available (as for the Purple Emperor). Such sites may constitute a vital resource for a species as a population's focal point in a complex environment extending over a considerable area and, because the butterflies may not actually breed at the site, access to it must be maintained by precluding imposition of any barriers to short-range dispersal. Many different species may utilize the same hill top, so that a number of hills are popular local collecting grounds. A 'hill' in this functional context need not be large, and hill-topping behaviour may be triggered by seemingly insignificant topographical relief (Baughman and Murphy, 1988).

Butterfly population structure is also a vital facet of attempting to define one of the most important terms used in discussions about conservation, namely 'rare'. 'Rarity' is a quality which appeals to decision-makers, and a rare species may readily evoke concern for its wellbeing. There are, though, several different ways in which the term is used. Consider, as examples, the following:

- A species which is widely distributed, continuously over a large area, but with high dispersion and thus a very low density. Very few individuals may be seen, or occur, at any one part of the species' broad range.
- A species which occurs disjunctly yet in a number of populations and localities. In this case the distribution may be very restricted, and the species will probably be absent from much habitat which appears to be suitable for it. It may, however, be reasonably abundant where it occurs.
- A species which is distributed as in the second example but with small populations or colonies.
- A species restricted to one site, where it is abundant.
- A species restricted to one site, where it occurs only in very low numbers.

Each of these situations could represent some interpretation of 'rarity', with either distribution, abundance or both used in the definition. Yet they are clearly different in ecological emphasis, and it may be very important in setting priorities to determine the precise status of a population or taxon. Any form of narrow geographical range (one site, few sites, small total habitat area) may give just cause for concern, and this may be especially so if only small numbers are present. In the next chapter, the factors leading to rarity and fragmented distributions are discussed further, but from the conservation aspect butterflies generally fall into one of three major groups:

1. Those which are common and widespread. They are not at present endangered in any sense as *species*, although particular local or outlying populations may from time to time be so. They need to be monitored to detect any sudden change of status, as a marked decrease in distribution or abundance, especially of ecologically sensitive or indicator taxa, may be an 'early warning' of the need for concern or action on a local or broader geographic scale. Many of these are useful 'barometers' of environmental health.

2. Those which are localized (in collector parlance, 'local') but generally abundant over the restricted areas in which they occur. A number of alpine taxa fall into this category. These distributions commonly reflect rather precise ecological needs, and such species may be very useful indicators of environmental quality because their wellbeing mirrors the overall health of local intricate environments or communities.
3. Those which are generally accepted as being 'rare' (however this is defined!) and which have perhaps contracted in range or abundance over historical time so that present colonies represent the remnant populations of species which were once more widespread.

Different subspecies of the same species may be referred to any of these categories, so that many butterfly species are ecological mosaics of widespread, local, and rare forms. The distinction between these groupings is therefore not always clear. The following Australian examples show the limited knowledge which is generally available as a basis for assessments, and some of the intricacies of the biology of particular species. They are given here to introduce the kinds of studies of individual species which need to be undertaken in order to detect significant features affecting butterfly populations.

The Common Brown (*Heteronympha merope merope*) is widespread in south-eastern mainland Australia, and two other subspecies (*H. m. salazar* and *H. m. doubledayi*) occur in Tasmania and southern Western Australia, respectively. As its name implies, this is an abundant butterfly in the south-east and is a familiar sight in grassy woodlands and open forest environments, especially at lower altitudes. It is univoltine, and adults typically start to appear in October or early November (Edwards, 1973; Fisher, 1978). During early summer males are usually much more common than females, but this gradually changes so that females predominate in late summer. At that time, males may be very scarce. Field samples taken over the flight season show that males rapidly become worn and 'tatty' after emergence.

Some females appear to emerge from the pupa well after most males have done so, while others emerge about the same time as males. Mating occurs very soon after the adults emerge, and males will pursue any female which flies by, sometimes mobbing her in groups of half a dozen or more. There is strong sexual dimorphism, with the female being particularly cryptic, looking like a dead leaf when at rest with her wings closed, but larger than the male and conspicuous while in flight. After mating, the females seek a sheltered locality and aestivate, resting through the heat of summer without maturing their eggs. Edwards (1973) did not find any females with developing ovaries or eggs before mid-February at low altitudes, but by late March all females that were dissected contained eggs. Even at this time, females appeared 'fresh' — a consequence of their relative inactivity. Females then commonly persist into May, whereas only a few very worn males do so. Males decline substantially in numbers from January onwards.

Aestivation here appears to be a counter to extreme climatic conditions at low altitudes and tends not to occur in the more equable regimes at higher altitudes. This means that young larvae are assured of a supply of young fresh food (various perennial grasses — Fisher, 1978, cited *Poa* spp., *Themeda australis*, *Brachypodium distachyon*) stimulated by late summer–autumn rains, and do not have to contend with tough dry grasses during the hottest times of the year. Eggs take about two weeks to hatch and larvae feed nocturnally from about May to September–October. Caterpillars shelter near the base of the food plants during the day, and are cryptic. Early instars are green, and a proportion of older caterpillars are brown. The pupa, which lasts for around five to six weeks, occurs on the ground.

Populations of *H. merope* are difficult to delimit, even though particular butterflies do not appear to disperse widely, because the apparent distribution can be across large areas with little evidence of breaks between segregates. However, the strategy of mating soon after emergence may reinforce local genetic identity, as opportunities for pre-mating dispersal are restricted and, apparently, females need to mate only once. Males will mate several times and stragglers in late summer will still pursue females emerging from aestivation. Mated females avoid males by 'diving' into the ground and remaining still, and their protracted aestivation until many males have died off may ensure that they are distracted only minimally when laying eggs. Both sexes feed freely from a variety of open flowers.

Despite having unique biological features in Australia, *H. merope* is of no current conservation concern. Its pattern of variation is clinal, both from north to south and from east to west (Pearse and Murray, 1982), and it has proved to be a valuable species to study in order to clarify some principles of butterfly evolution in Australia.

The Alpine Silver Xenica (*Oreixenica latialis*) is one of several alpine Satyrinae in south-eastern Australia, and is restricted to a number of isolated highland areas (mostly well above 1000 metres) in Victoria and New South Wales. Three subspecies have been named: *O. l. latialis* is widespread and often abundant where it occurs above about 1500 metres, *O. l. theddora* occurs only on Mount Buffalo (Victoria), and *O. l. nama* is a variable form from around Nimmitabel (New South Wales) and possibly intergrades with *O. l. latialis*. As Common and Waterhouse (1981) stated, closer study is required to determine the status of these populations. The univoltine butterfly appears in February and March, sometimes extending to April, and replaces the earlier flying *O. orichora* on alpine grasslands. Adults do not move far, at the most a few hundred metres, but populations extend over hundreds of hectares. Eggs (of *O. l. latialis*) are green and deposited singly or in small groups on grasses.

Caterpillars are initially green but late instars tend to be brown; they feed on native *Poa* tussock grasses. The brown pupa lies on the ground amongst the tussocks. It appears that both egg and pupal stages are of relatively short duration, and the caterpillars overwinter deep in the tussocks.

Despite its very restricted distribution, it is doubtful if *O. latialis* is threatened, but the increasing pressures of tourism (particularly of accommodation for winter sports) on some of the areas where it occurs has some potential to increase vulnerability, as does destruction and trampling of native vegetation by cattle. It has the potential to be an 'umbrella species', whose wellbeing mirrors that of a host of less conspicuous invertebrates in sensitive habitats.

The Swordgrass Brown and the Australian Hairstreak. Two species are discussed here, both of which have developed series of localized subspecies in Australia, but which exemplify rather different concerns for their long-term conservation. **The Swordgrass Brown** (*Tisiphone abeona*) has long aroused interest as a complex of forms around the coastal areas of south-east mainland Australia. Seven 'subspecies' have been named and a further restricted and variable population known as *T. a. joanna* is thought to be maintained by hybridization between two of these. The complex variation of this species, possibly one of the more evident examples of current speciation in the Australian butterflies, has been studied by several workers — Waterhouse (1928), for instance, undertook artificial hybridization trials between various forms, and was responsible for clarifying the status of *joanna* as a hybrid restricted to a small zone where two subspecies meet. Although most of the subspecies are not individually endangered or particularly rare where they occur, each occupies a rather small part of the total range of the species, and each merits consideration as part of an evolutionary radiation worthy of further study. Most are geographically isolated from each other and some may occupy only a small area.

T. abeona is restricted to swampy areas, where the larvae feed on sword grass (Cyperaceae: *Gahnia* spp., with several species recorded as suitable foods), and destruction of swamps on an increasing scale is threatening the survival of *T. a. joanna* in New South Wales. Various populations of most other forms are also highly vulnerable. Males sometimes seem to be more common than females, but both sexes tend to be furtive and not to fly far from the swamps — or sometimes gullies in open woodland — in which the food plants grow. The populations are therefore discrete and usually occupy only small areas. Most subspecies apparently have two generations a year, with adults appearing in spring and again in late summer.

Eggs are laid singly on *Gahnia*, and Common and Waterhouse (1981) summarized the following life cycle for *T. a. abeona* near Sydney. Eggs hatch in 9 or 10 days (December)–16 days (October, March). Larvae from eggs laid in spring persist for three to five months and those from autumn eggs take five to eight months to develop. The pupal stage takes 24–45 days (August–September) but less for the second generation (19–26 days October–November; 18 days December–February). Larvae hide head-downwards between leaves in the centre of dense *Gahnia* clumps during the day and feed in the early morning, early evening, or at night. The pupa is suspended head downwards, and commonly from the pendant outer foliage of the food plant. All early stages are green.

The Australian Hairstreak (*Pseudalmenus chlorinda*) is an endemic lycaenid. Like *T. abeona*, it shows strong geographical variation which has resulted in designation of seven subspecies in south-east mainland Australia and Tasmania. The distribution of some is very restricted, and only *P. c. zephyrus* is at all widespread. The others are regarded as rare and local and several are known from only single localities, a situation not uncommon for Lycaenidae in Australia and one which gives such taxa high conservation status. In Tasmania, as Common and Waterhouse (1981) pointed out, populations separated by only a few kilometres differ substantially in markings. Several subspecies have been placed on various 'priority' lists for conservation.

Populations or colonies of *P. chlorinda* are often small and adults are not known to fly far. Adults occur in spring, and the species is generally univoltine. Eggs are laid singly or in small groups on *Acacia* twigs or stems, and the bipinnate *A. dealbata* is the preferred (or sole recorded) food plant of some subspecies, although *P. c. zephyrus* has been recorded from phyllodinous wattles (such as *A. melanoxylon*) in New South Wales and near Canberra. Both small (from a few centimetres high) and large (5 metres or more) hosts may be selected.

Larvae are tended by small black *Iridomyrmex* ants and are fully grown within about a month at lower altitudes. Pupae are sometimes also tended by ants: pupation takes place under bark, often of neighbouring eucalypts several metres from the *Acacia*. There is some suggestion of a partial second generation in some subspecies, but most populations seem to be truly univoltine. The pupae do not hatch for around eight to nine months, in late August–October.

The precise status of some populations or subspecies of *P. chlorinda* is by no means clear, although there is little doubt that some are indeed very restricted. Couchman and Couchman (1977) emphasized that it is highly vulnerable to habitat change and has suffered substantially in Tasmania, where it has been regarded as endangered. As the butterfly needs the close association of an *Acacia*, an attendant ant and a eucalypt (usually *E. viminalis*), it has been eliminated over considerable areas where one or more of these resources has been destroyed by land clearing. The Couchmans noted that they found *P. chlorinda* in more than 50 Tasmanian localities in the years after 1945 but that most of these had been destroyed, so that by 1977 it was 'difficult to think of 10 areas within the island where the Hairstreak may survive'. Such practices as pasture improvement (with removal of mature eucalypts and wattles) eliminated *P. c. conara*, and clear felling associated with the woodchip industry resulted in destruction of prime habitat of *P. c. chlorinda*. Two (unnamed) forms are now extinct, and the Couchmans stressed the urgent need for reserves in order for the butterfly to survive in Tasmania.

It is commonly observed that butterflies found in small scattered colonies, like *T. abeona* and *P. chlorinda*, are absent from many areas where their food plants are common and which appear to be thoroughly suitable habitats, though a metapopulation structure (p. 33) may be a partial explanation in some cases. This form of rarity is very difficult to understand, and has long intrigued biologists. Human disturbance of the habitat is sometimes invoked as an explanation for such distributions, but this is usually only part of the story, if it is involved at all. Cappucino and Kareiva (1985) have studied the biology of one North American woodland butterfly of this sort, the West Virginia White (*Pieris virginiensis*). Colonies of this species had been observed to become extinct for no apparent reason, and no single factor was found to be responsible for the species' rarity. Its reproduction was hindered in different ways throughout its development. Initially, in spring (April) short-lived adults emerged from overwintering pupae in unpredictable weather. In one year of the study, about 70% of the spring weather was unsuitable for flight, so that the butterflies may have had little opportunity to lay eggs. Even if they can disperse, females searching for *Dentaria* plants on which to oviposit have difficulty doing so if these are overgrown by other plants. Females which do encounter a good patch of food plant may fly straight on rather than change to a more intensive form of searching behaviour that would ensure that they would remain in the neighbourhood of these. However, adults do not readily cross open fields, and one implication of this is that if a colony becomes extinct recolonization is unlikely to occur unless there is another population within the same woodland; if not, the patchy distribution may be enhanced.

In contrast, the Black Swallowtail *Papilio polyxenes* in Costa Rica feeds on a very patchy food plant (the umbellifer *Spananthe paniculata*), but is able to live in this heterogenous environment because its greater dispersal ability enables females to colonize new patches of the short-lived plant as they appear (Blau, 1980). New food patches thus provide a continuous sequence of refuges from predators and parasites, and from competition resulting from high butterfly populations. For *P. polyxenes*, food plant patches are produced by localized disturbances in an environment where the growing season is continuous. Emerging female butterflies disperse to reach new habitat patches, but some lay eggs before leaving so that the species can 'capitalize' on any remaining food, and may undergo one or two additional generations in the same place. The patch degenerates naturally after about six months because of succession and senescence. This butterfly is a good example of one dependent on a continuous sequence of short-lived temporary habitats over a broad distributional range.

In the case of the West Virginia White, as the early stages develop, caterpillars after instar IV may have to move to find a second food plant, as the original one would have been eaten or become senescent and unsuitable by then. However, they apparently are attracted just as readily towards erect non-hosts as to *Dentaria*, so that searching in low-density areas may be largely futile. Not only may these caterpillars starve through wandering away from patches of food plant, but while on the ground they may be eaten by spiders, ground-beetles and others. Interestingly, the likely dangers to larvae are not minimized by the laying mother females, which do not lay preferentially either on larger food plants (where the greater amount of food could minimize the need for later movement) or on dense clumps of *Dentaria* (where searching larvae could more easily find new food). Unpredictable weather can also alter the seasonal appearance of the butterfly in relation to its food plant, because *P. virginiensis* must complete its life cycle within a narrow 'time window', after winter cold and before host plants senesce. If butterfly emergence is retarded by even a few days, food plants may be much less available. In at least one area, *P. virginiensis* utilizes a second host, *Arabis laevigata*, which is less ephemeral than *Dentaria* and seems to be preferred when both plants are present (Shuey and Peacock, 1989). This sort of need for close seasonal regulation of the life cycle also occurs in some other Pieridae and members of other families. In the well-studied nymphalid *Euphydryas editha*, late-hatching larvae may not be able to complete development before senescence of their food plant, *Plantago erecta*.

BEHAVIOUR

From this brief survey of some aspects of butterfly populations and life cycles, we can infer that many species are restricted in their resource and habitat needs and that even very closely related species may differ substantially in their biology and behaviour, or that the needs and developmental pattern of a single species differ over a broad geographical range. It is therefore very difficult — and unwise — to extrapolate uncritically from knowledge of one species to assess another when planning conservation measures. Indeed, even variability within a species may make it foolish to generalize about the ecology of that species from ecological data derived from a limited sample of populations of that same species (Ehrlich *et al.*, 1980). All too often, the information needed to assess any given case is simply not available. There is a great need to increase our general foundation of knowledge of butterfly biology and to determine what generalizations can be made. Even most common species have not been adequately documented, and differing local conditions may make many of these much more variable than is usually suspected. Finding out 'what butterflies do' and 'why they occur where they do', as well as answering such seemingly naive questions as 'How far can or do they fly?', 'What do they eat?' and 'How do they find mates?' can be extraordinarily challenging. Some of the problems involved will be addressed in Chapter 5, but all these questions (and others) are ones which amateur naturalists can help to solve by watching butterflies and recording their observations. In his fascinating book *Butterfly Watching*, Whalley (1980) emphasized the value of even very basic observations — such as the times of day butterflies are active, how this is affected by weather, and whether they select particular kinds of flowers for feeding or laying on — and made the very pertinent comment that 'As a principle, no observation is too insignificant to be noted down; it may be the very feature which is the key to an unanswered question'. Whalley was writing predominantly on the much studied butterfly fauna of the United Kingdom, and his point is even more relevant in relation to the butterflies of the rest of the world.

Many such observations are significant in considering ecological associations and segregation — essentially, we need to know the 'key' resources needed by a given butterfly species, and how these may be provided in order to conserve it. We shall see later that lack of such knowledge has cost some species dearly as terrestrial habitats become altered by human 'progress'. Earlier in this chapter, I noted some aspects of butterfly behaviour, such as their dispersal, selection of oviposition site and feeding, but others are also very fruitful fields for study and observation. It is clearly not possible to provide a total picture here, but the following are noted as additional examples.

Thermoregulatory behaviour (see also p. 14). In his *Natural History of Butterflies*, Feltwell (1986) noted that the diurnal activity rhythms of butterflies are often related to sunlight (heat) and wind. Thermoregulatory behaviour (by which butterflies elevate their body temperature to the range, commonly around 28–40°C, needed for flight: Kingsolver, 1985) is often complex and involves a range of postures and activities. Temperature is a major influence on butterfly activity.

Clench (1966) was among the first to categorize some of the behaviour involved. His 'primary heat gain devices' included:

- Dorsal basking — the butterfly rests in the sun, opens its wings fully, and orientates itself so that the upper surfaces are insolated to the greatest possible extent. Although this strategy is widespread, some butterflies apparently do not indulge in it at all.
- Body basking — the body is orientated to the sun with the wings only partially open. This may be rather restricted and unusual.
- Ground contact — the butterfly rests on sun-heated ground and presses its body close to the substrate.
- 'Shivering' — the generation of metabolic heat by rapid wing vibration. This habit is much more common in moths than butterflies but has been observed in some skippers.

- Avoidance of wind is likely to enhance the effectiveness of most of the above by reducing conductive loss of heat to the air.

There are also 'primary heat loss devices', such as hyperventilation and seeking shade — the latter being a passive form of heat control. Clench also distinguished several different behaviour patterns related to thermoregulation in butterflies, and one or other of these occurs throughout adult life. Observations on posture and daily activity rhythms of butterflies may be very pertinent in assessing a butterfly's thermal requirements — when it is avoiding heat, becoming active, 'seeking' to prolong activity by warming early or late in the day, whether the two sexes differ in activity or colouration (often a correlate of thermoregulatory efficiency), whether postures or resting sites change throughout the day, and so on.

Territorial behaviour. Some form of territorial behaviour seems to be more common in butterflies than has been generally realized, and in a review of insect territoriality Fitzpatrick and Wellington (1983) commented that 'more ornithologists than entomologists' had noted the territorial behaviour of male butterflies. Many butterflies if disturbed will return to the flower or other perch on which they were resting, and males of some species leave such vantage points to actively chase away others of the same species and may act aggressively towards many moving objects. There are even records of butterflies attacking people and various birds. When two males of the same species meet they may climb steeply and turn tightly, a traditional 'aerial combat' manoeuvre (Baker, 1972). The wings of territorial males may become very ragged through being used to flail other butterflies (Wellington, 1974).

In Britain, the woodland satyrine *Pararge aegeria* (the Speckled Wood) defends patches of sunlight in woodland. A male will 'adopt' such a patch as his territory, seemingly a mate-seeking device akin to hill-topping (p. 34) as females are also attracted to sunlight, and chase off other males. Males of some other Nymphalidae establish territories at potential oviposition sites or along likely routes to these. *Heliconius* males patrol and defend 10–15 metre wide sunny corridors against conspecifics, and if they leave their territory to do this the area may be taken over by a vagrant male. However the primary resident (in all of 149 encounters viewed) could evict such newcomers from a territory left only temporarily vacant (Benson *et al.*, 1989).

Some skipper males seem to be among the most aggressive defenders of territories. The neotropical *Hamadryas* (Nymphalidae) 'warns off' other males by producing a loud staccato 'crashing' noise with its wings.

In quite a lot of butterflies, the males appear before the females, a situation known as 'protandry', so that the sex ratio in the population changes substantially during the flight season. This trait is often linked with territoriality, and males may have established a territory even before females appear in the population.

Courtship and mating. Territoriality is an important facet of mate-seeking by male butterflies. By emerging earlier than females they can also maximize their chances of meeting virgin mates, especially in those species in which females may mate only once (Wiklund and Fagerstrom, 1977). Protandry is, of course, largely restricted to species with discrete generations and which are essentially seasonal in appearance, often with a very well-defined seasonal flight pattern. Males of a few butterflies actually aggregate and wait for females to emerge. Males of the common Australian lycaenid *Jalmenus evagoras* (the Imperial Blue) emerge slightly before the females. They crowd around female pupae on *Acacia* trees and actively jostle with each other to determine which one will mate as soon as the female emerges (Pierce, 1984).

In general there are two seasonally well-defined strategies which male butterflies use to find mates — on the assumption that, most commonly, the males initiate the choosing behaviour (see discussion by Silberglied, 1984). There is no doubt that the bright colouring of so many butterflies plays a role in mate recognition and courtship, although sexual scents (pheromones) are now also recognized as important in some groups. In some butterflies the more important role of bright colours, especially in sexually dimorphic forms with bright males, may be for recognition of other males in the overall context of mate location behaviour. Butterflies certainly have excellent colour vision. Females may accept male advances, or reject them by characteristic changes in behaviour, as we noted for *Heteronympha merope* (p. 34). Male butterflies may be either 'perchers' or 'patrollers'. The strategy adopted is sometimes very characteristic of a given taxon, but some species may adopt either one. Male Speckled Woods, for instance, alternate the two strategies (Shreeve, 1984). Patrolling males of this species locate females faster, but flights are related to the need to maintain high body temperatures. Both strategies may involve territory, as a male then either perches in 'his' area and flies to inspect any likely mate entering it, or patrols the area more continuously with the same objective. These patrolling species often tend to have larger territories than habitual perchers, and the 'beats' of several males may overlap.

Courtship behaviour, especially that involving attraction to colours or shapes, can be studied experimentally to determine particular important cues. Collectors have long known that a dead butterfly exposed in the flight path of others can act as a lure, and a number of major studies of butterfly courtship have employed 'tethered' or 'harnessed' live individuals. Brower (1959) tethered female swallowtails of the *Papilio glaucus* group on fine thread attached to a long horizontal wire suspended across the flight path of males. Interspecific attractions can occur, and males of some species can be attracted to rather crude artificial models of female butterflies. *Papilio ulysses*, for example, can be lured to pieces of metallic blue foil or paper placed on the ground where they are flying.

Such studies have been very informative in determining the stimuli used by male butterflies to approach females.

We noted earlier that chemical communication may also be important. Males of many butterflies have various forms of scent-producing 'androconia', commonly as patches of modified scales on the wings. The male of the Grayling (*Hipparchia semele*: Satyrinae) can stimulate a non-receptive female by bringing her antennae into contact with his forewing androconial scales (Tinbergen, 1942). Danaine males have extrusible abdominal 'hair-pencils' which play a vital role in courtship. A classic study of courtship of the Queen butterfly (*Danaus gilippus*) by Brower *et al.* (1965) showed how the male's hairpencils were used to disseminate scent near the female as integral phases of successful courtship. Such courtship sequences (Figure 2.6, Table 2.2) are a fascinating and revealing aspect of butterfly behaviour. The ritual behaviours revealed by such studies show clearly a sequence of interplay between various forms of visual, chemical and tactile stimuli which have developed to produce highly specific mating behaviour which can function as an effective isolating mechanism for closely related species.

A courtship sequence may have different pathways. The Orangetip (*Anthocaris cardamines*: Pieridae) in Europe, for example, can show three sequences depending on whether the male finds the female in flight or perched, and also whether the perched female then takes flight (Figure 2.7; Wiklund and Forsberg, 1985).

Studies of butterflies' 'time budgets' can tell us much about how their territorial and mating systems work. A study of a population of a Canadian satyrine, *Oeneis chryxus*, over three years by Knapston (1985) showed that males defended territories in forest clearings by chases and spiral flights with conspecifics. They generally perched on the ground, more rarely on vegetation, and took short flights within and around an area, probably patrolling the boundaries of a territory (Table 2.3). If another butterfly was encountered, a chase or spiral flight ensued rapidly — all butterflies chased were males, and the chase distance was for up to about five metres. Spiral flights, although usually involving two males, occasionally contained up to four! Butterflies sometimes flew to a height of 50 metres, and these flights frequently took males outside their territories. Most returned to their territory directly or with a short delay after the encounter finished. Such 'spiral flights' are apparently common in Nymphalidae.

Figure 2.8 (p. 48) shows the pattern for the Red Admiral (*Vanessa atalanta*) in which males establish and defend territories only for about 2 to $2\frac{1}{2}$ hours in late afternoon (Bitzer and Shaw, 1980). Males of *O. chryxus* did not collectively occupy all suitable habitat, and their territories contained neither food nor oviposition sites, yet each male defended a given patch of ground for some time — in some cases, up to 11 days. Territories varied somewhat from day to day

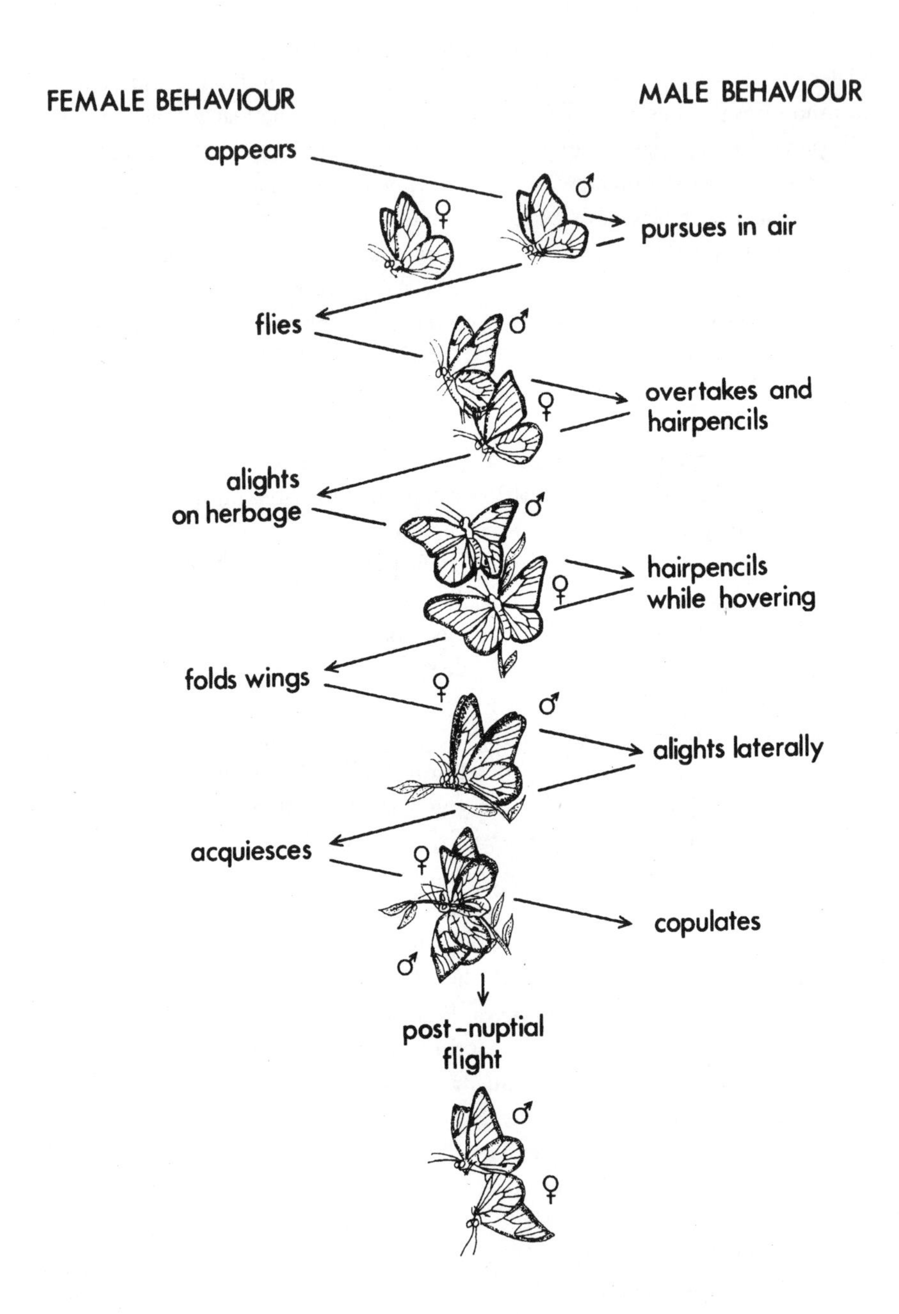

Figure 2.6 The courtship behaviour of the Queen butterfly, *Danaus gilippus* (Brower *et al.*, 1965).

Table 2.2 Components and phases of sexual behaviour in the Queen Butterfly, *Danaus gilippus*. See Figure 2.6 (Brower *et al.*, 1965).

I First aerial component	
	PHASE 1 Aerial pursuit
	PHASE 2 Aerial hairpencilling
II First ground component	
	PHASE 3 Ground hairpencilling
	PHASE 4 Hovering and striking
	PHASE 5 Copulation attempt
	PHASE 6 Copulation
III Second aerial component	
	PHASE 7 Post-nuptial flight
IV Second ground component	
	PHASE 8 Insemination
	PHASE 9 Termination of copulation

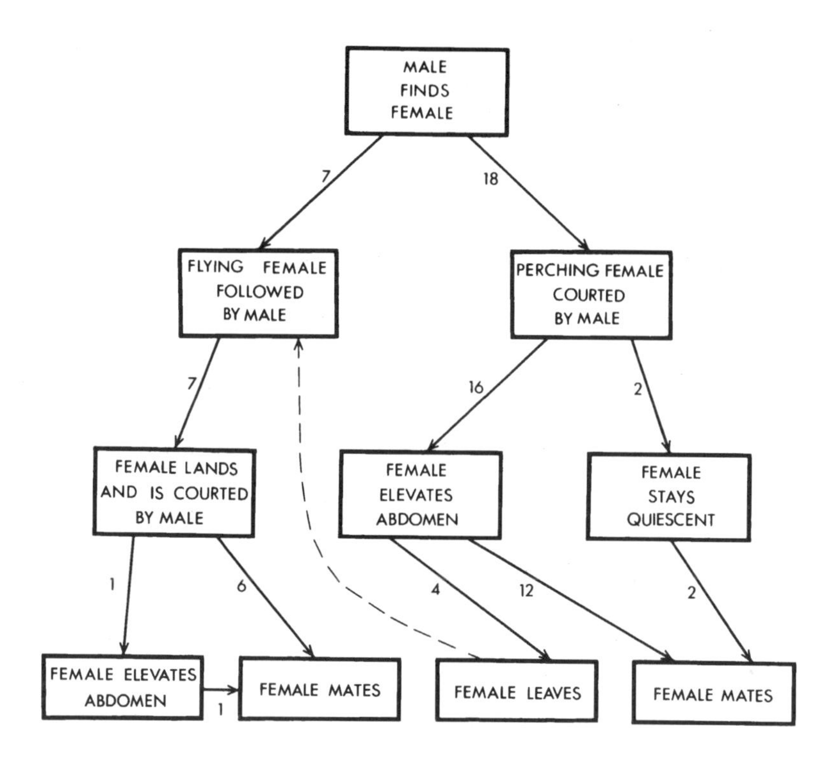

Figure 2.7 The mating sequence of the Orangetip, *Anthocaris cardamines* (Pieridae), showing alternative behaviours with (as figures) the number of instances observed for each (Wiklund and Forsberg, 1985).

Table 2.3 Time budget analysis of observed behaviour in male Chryxus Arctic butterflies, *Oeneis chryxus* (Knapton, 1985).

	1982	*1983*	*1984*
Total observation time (h)	32	42	38
Percentage time spent			
Resting/perching	79.1	74.3	77.2
Short flights	12.9	15.3	8.4
Intraspecific and interspecific interactions	7.1	9.2	12.3
Feeding	0.9	1.2	2.1

Note: These values represent the percentage of total time for all butterflies within each year in each behaviour category.

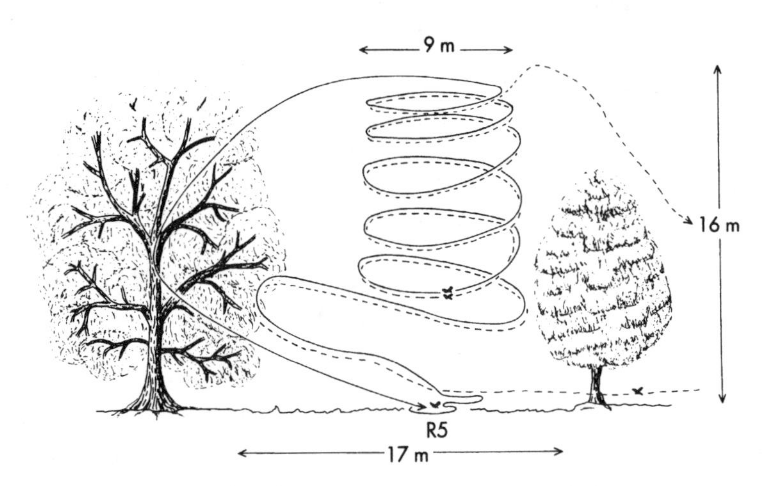

Figure 2.8 The territorial interaction of male Red Admirals, *Vanessa atalanta* (Nymphalidae). The 17 m ground arrow indicates the territory boundary; RS, resting spot of resident; solid line is flight path of the resident male and the dotted line is that of the intruder (Bitzer and Shaw, 1980).

(Figure 2.9) but did not undergo major rearrangements. Knapston (1985) considered that *O. chryxus* more resembled some vertebrate 'lek display' systems (such as occur in some birds where males gather to display to females) rather than more conventional territoriality allied to hill-topping or other landmark-based butterfly territoriality. Unlike birds and most other groups of animals in which territory has been studied, a butterfly's territory may be transient; some species may establish two or more territories in different places in a single day as sunlight patterns and direction change from morning to evening.

Seasonality and life history patterns. We noted earlier that most temperate region butterflies undergo discrete generations each year, and in tropical areas many butterflies are probably far more seasonal than is commonly supposed (Spitzer, 1983). However, caution is needed in interpreting life cycles even of temperate butterflies, as the basic pattern of development may be modified substantially by local conditions. The Speckled Wood in Sweden, is one example. It is bivoltine in the south of the country, but univoltine further north, paralleling the Common Blue in the UK (p. 21). Larvae of the northern ('central Sweden') population normally aestivate for part of the summer, whereas southern ones develop directly. These are alternative strategies, and the balance between them apparently differs in the two populations (Wiklund *et al.*, 1983).

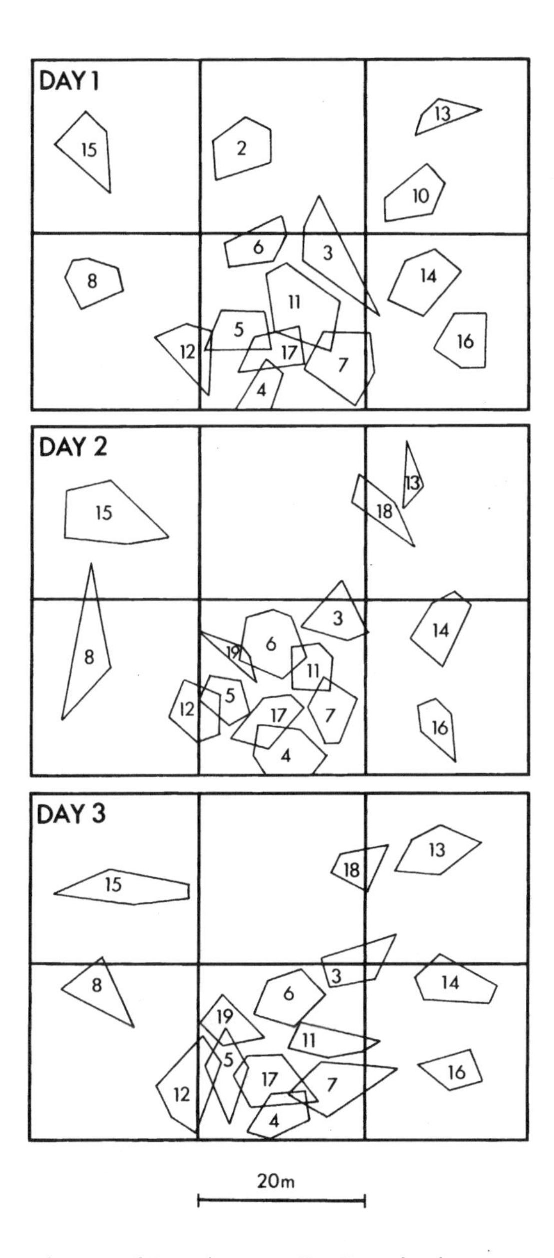

Figure 2.9 Distributions of territories of *Oeneis chryxus* on a 60 × 40 m grid on three consecutive days. The numbers correspond to the territories of particular marked butterflies (Knapton, 1985).

CHAPTER THREE

CAUSES OF BUTTERFLY DECLINE

All insect species vary in abundance because of differences in birth rate and death rate, which are influenced by changing environmental factors. Each may have a characteristic long-term abundance, so that we can say with confidence that a particular species is 'common' (abundance high) or 'rare' (abundance low), but undergo considerable variation about this equilibrium level from generation to generation. For example, if adult food supplies are good (abundant flowers producing nutritious nectar for butterflies) their reproductive output (number of fertile eggs laid and offspring produced) may be raised; if larval foods are then abundant and predators and parasites scarce, subsequent caterpillar survival may be high, and warm temperatures might hasten the development of all stages. The converse also applies: poor food supplies may result in few eggs and/or low larval survival or small stunted adults, large numbers of predators and/or parasites may take a heavy toll of immature stages, and bad weather may retard development or cause death more directly. In essence, the supply of resources, and the effects of both the physical environment and of other organisms present in the ecological community, inevitably impose variations in the numbers of any butterfly species, and in the amount of change between any two successive generations. Such changes can result in a several-fold variation in abundance in succeeding generations. Likewise, the extent of migration of some species may differ from year to year — collectors are very familiar with 'good' and 'bad' years for migrant butterflies. Records of migrant species, especially those on the outer edges of their usual dispersal range, show that many species vary considerably in their incidence there.

Butterflies are renowned for their large fluctuations in abundance from year to year (Warren, 1992). As well as butterfly numbers varying continually, the habitat where any butterfly species lives is also dynamic and changing all the time. At some times it may be able to support more individuals (that is, have a higher 'carrying capacity') than at others. Later, we shall establish that 'habitat' is the key resource for butterflies. It is now widely recognised that the life history of a species — its 'ecological strategy' — which incorporates many of the themes of the preceding chapter, reflects features of the habitat: what Southwood (1977) has called the 'habitat templet'. As Southwood noted, three parameters are particularly important. These, in very general terms, are:

- the length of time that a habitat remains suitable for a given species;
- the extent to which carrying capacity varies because of fluctuations in vital resources, such as food plants;
- the degree of spatial variation or 'patchiness' — that is, whether the whole habitat is relatively uniform or with only parts of the area being so. This may reflect both major physical/vegetational differences and small differences in aspect or topography influencing local microclimates, so that a metapopulation pattern may be favoured over a more continuous occupation of the habitat.

Habitats change. Specialist species — those which need particular resources in order to thrive — must either move in space to track those resources, or adjust their life cycles in time so that, for example, the incidence of caterpillars coincides with the presence of a seasonal food plant. Other than short-term regular (and therefore predictable) seasonal changes, the most important natural change is due to 'succession'. This is the process by which open areas such as grassland are gradually invaded by a variety of larger plants so that over many years they are transformed into shrubland and, eventually, woodland or forest. Many of the early stages in a succession occur very rapidly, so that each does not last for long, and these changes may include subtle replacement of some plant species by others and not, necessarily, major changes visible to the onlooker. Later stages are slower, so that a woodland or forest is dominated by long-lived trees and changes rather little during a human lifespan.

Shade in woodland rides increases as the trees mature, and regulation of shade may be important in the management of some woodland butterflies, perhaps predominantly because of its effects on food plants. Many plant species are adversely affected by increased shading, whether this is caused by natural maturation (UK: Sparks *et al.*, 1996) or replacement with plantation trees (as for *Luehdorfia japonica* food plants in Japan: Ishii, 1996). *Luehdorfia* oviposits on perennial *Heterotropa* herbs, which thrive in well-lit coppices but not in areas shaded with introduced conifers. A continued increase in shading is associated with a decline in suitable resources in the UK, and is likely to lead to local extinctions of some butterfly species (Sparks *et al.*, 1996).

Many insects, including butterflies, which depend on early successional habitats may not be able to persist in the same place for more than a very few generations, because it has by then become unsuitable in some way. In contrast, a woodland butterfly population may be able to persist for a long time in one place, so that populations may be relatively sedentary. A high level of dispersal (migration) is commonly associated with a need to keep pace with changes in habitats and to find new breeding areas at frequent intervals.

Decline in butterflies is widespread, and has been detected amongst almost all faunas which have been appraised. In Japan, none of the 240 species has yet become extinct (Ishii, 1996), but local populations are 'degraded or threatened everywhere'. Forty-seven taxa are of special conservation concern, and the two 'endangered' species — *Shijimia moorei* (Lycaenidae) and *Fabriciana nerippe* (Nymphalidae) — occupy different habitats and illustrate the problems of habitat change in that country. *S. moorei* is confined to two mountainous districts, where its larvae feed on a species of Gesneriaceae (*Lysionotus pauciflorus*) which grows on the mossy trunks of old evergreen oaks in climax forest. The butterfly is threatened by the deforestation of these mature systems. Grassland species, including *F. nerippe*, have declined widely as a result of the loss of early successional stages formerly maintained by traditional farming. The food plant of *F. nerippe*, *Viola mandshurica*, grew abundantly in such grasslands, but did not thrive in longer swards: the butterfly has thus undergone considerable range contraction and is now confined to small parts of western Japan (Figure 3.1).

As in the European Alps (Erhardt, 1995), butterfly decline may be most pronounced in lowland areas subject to greater human disturbance. In Switzerland, Erhardt noted that 20 out of 57 lowland species (35%) were highly threatened, whereas this proportion declined at higher altitudes: lowland to subalpine (8 of 80: 10%) and subalpine to alpine (1 of 36: 2.7%). Intensification of agriculture is the major threat to butterflies in the Swiss Alps (Figure 3.2), but a wide range of processes contribute to their overall decline. All the effects serve to reduce and fragment suitable habitat, and some (particularly those associated with tourism infrastructure, clearing for ski-runs, and increases in other sports) are largely restricted to these regions in the severity of their impacts on such sensitive systems. Similarly, Larsen (1995b) regarded small areas of montane forest as 'probably the most threatened African habitats'.

If the habitat changes to the extent that resources vital for the survival of a butterfly species are no longer available, the options for the butterfly are to move to a different area where the habitat is suitable, to change habits so that it can exploit different resources in the same place, or to become extinct. The ability to change is often limited, although we have seen that some degree of ecological flexibility is relatively common. In general terms, many species in nature have some potential to adapt to changes in their environment, but this option is becoming progressively restricted. One, and perhaps the most important, ramification of human changes to so many natural habitats for our own convenience is that these are both more sudden and more drastic than many naturally occurring changes. In closing the option of gradual adaptation, extinction

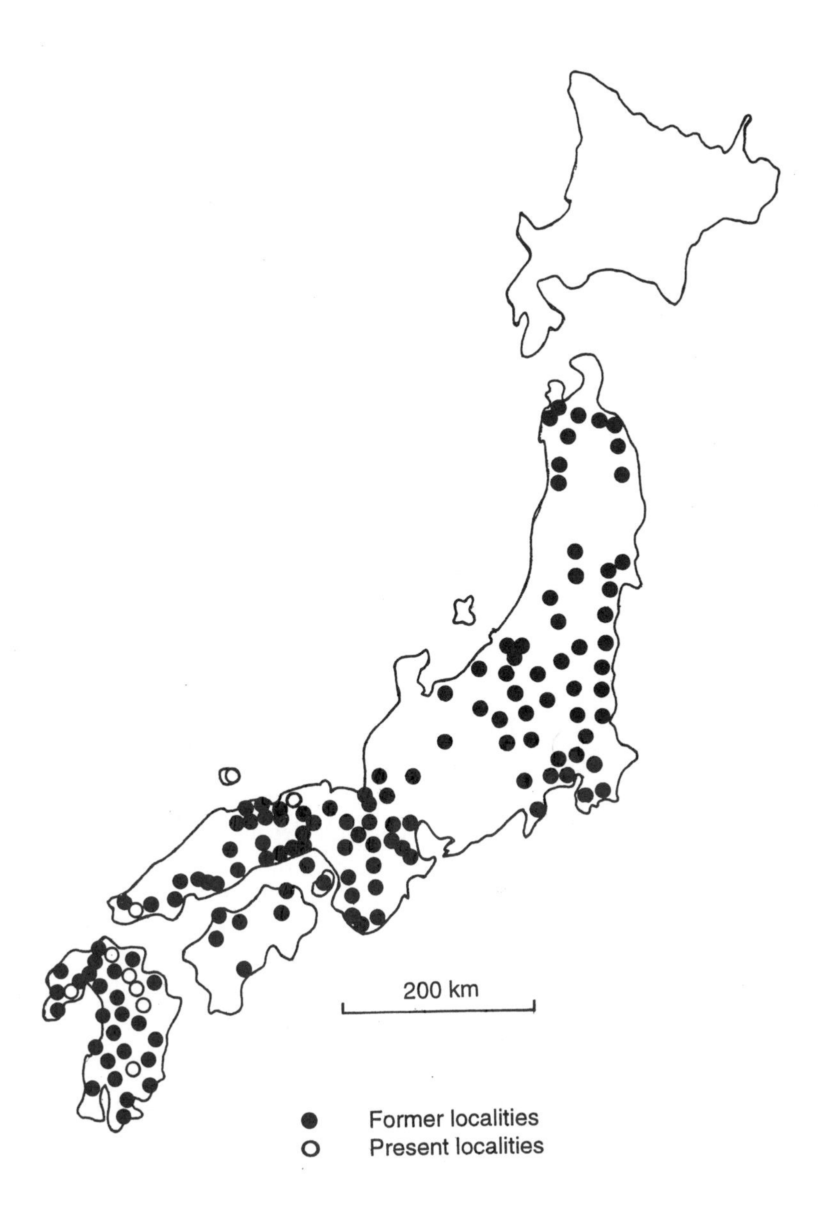

Figure 3.1 Distribution of *Fabriciana nerippe* in southern Japan (after Ishii, 1996). Closed circles indicate past localities, open circles indicate present localities.

becomes the major option remaining for many animal and plant species. Habitat patchiness means that local populations may be able to persist only if immigration from other patches is feasible in order to counter local extinctions (Morton, 1983). Nevertheless, many butterflies have adapted to anthropogenic

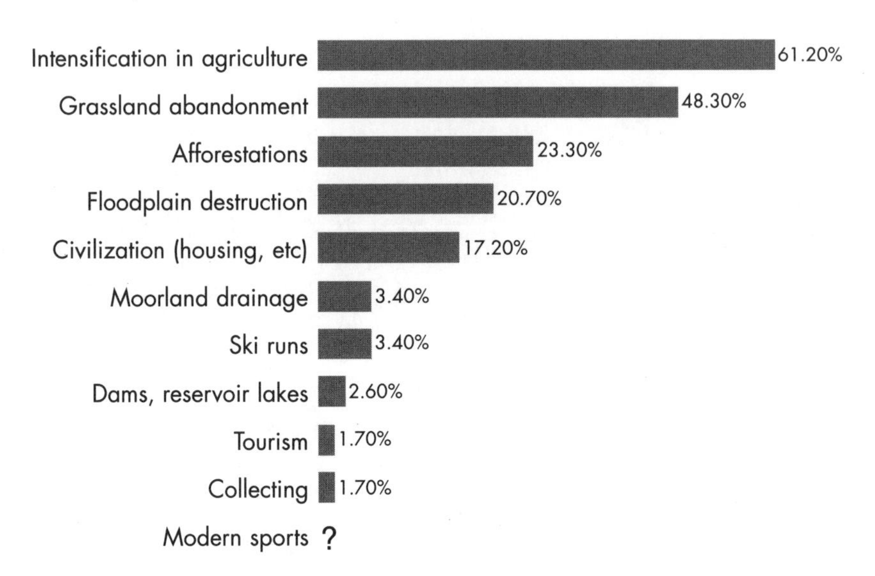

Figure 3.2 Categories of threat to alpine butterflies in Switzerland, with relative proportions of taxa affected (n = 116 species, Hesperiidae excluded) (after Erhardt, 1995).

landscapes, and many habitats that we tend to regard as 'natural' are in fact the products of long periods of human management. Traditional farming and forestry practices, for example, are important determinants of butterfly habitat quality in much of Europe.

Small patches of primary rainforest retain only depauperate butterfly faunas in relation to larger nearby patches (Daily and Ehrlich, 1995). Habitat fragmentation increases the likelihood of extinction of butterfly species because:

- populations may not be able to survive in small areas,
- there may be genetic problems associated with inbreeding and reduced gene flow, and
- recolonization may not occur following local extinctions (Warren, 1992).

However, the causes of decline are likely to vary over the range of a species, and — except for the direct loss or degradation of habitat being of paramount concern — may be very difficult to determine. The extent of loss which has already occurred is documented only for a few well-known faunas. In the Netherlands, 15 of 71 species have become extinct, representing one of the highest recorded levels of extirpation, and another 36 species have declined (van Swaay, 1990, 1994). Attention was drawn to the situation in 1993 with the release of a postage stamp in a miniature sheet on which the problem species were listed by name. The levels of 'endangerment' (rather then documented extinctions) are high in several other European countries for which some information is available (Switzerland 39%, Germany and Austria about 40–50%:

Erhardt, 1995), but on a more local scale the levels can be much higher. Thus, in the Seeland area of Switzerland, 53% of species are extinct and 27% are endangered, so that only 20% are not threatened (Bryner, 1987, quoted by Erhardt, 1995).

A somewhat different approach to assessing butterfly declines in the UK involved looking at changes in species richness at a series of 'hotspots' (centres of species richness) between the periods 1930–1970 and 1970–1983, based on the 10 km square grid used for mapping butterfly distributions (p. 125). 'Hotspots' are the top 5% of recorded squares ranked by number of species: the first period has 218 hotspots, the second 118 hotspots. Between the two periods, 134 hotspots were lost (that is, they fell below the defined threshold of 31 species), 84 were maintained, and 34 were gained. Despite the need for caution in interpreting the information available, the widespread decreases in butterfly richness appeared to reflect loss of habitat patches and increasing isolation (Prendergast and Eversham, 1993).

Change is usual, and a common problem in conservation is trying to assess what degree of numerical fluctuations or habitat alteration is natural, to establish the background 'noise' against which to appraise more drastic effects brought about by human activities. Despite implications from the framework given in the last chapter, the ecological needs of most butterflies are *not* well understood and, although we can sometimes grossly categorize the natural influences on their numbers and the relative extent or importance of these, sound quantitative information is usually not available. Heath (1981) noted that what is happening now is in some instances the end point of a long process: a decline in numbers of a butterfly from five million to 50 000 individuals may go unnoticed, but a decline from 50 000 to 500 (or less) will often be much more obvious.

Beirne's (1955) survey of causes of fluctuations of British Lepidoptera (including moths) since the 1860s implicated 'natural enemies' (a collective term for predators, parasites and disease) as a major influence, as well as weather effects. Most changes from 'bad' to 'good' years were readily associated with weather extremes or striking weather conditions, but the converse was not always true. Beirne suggested that natural enemies, rather than weather conditions alone, may have been primarily responsible for fluctuations: build-up of numbers of Lepidoptera in a 'good' year encourages large populations of, especially, parasitic wasps which then reduce Lepidoptera abundance later. The main factors contributing to good years included severe cold in the previous winter, with consequent adverse effects on the natural enemies, and warm dry weather in summer, promoting rapid development and adult activity and reducing incidence of diseases. Factors contributing to bad years were good years immediately preceding, so that natural enemy populations were high, relatively high rainfall over several consecutive years, which appeared to promote the incidence of diseases, and effects of adverse weather such as droughts or late frosts (which could also kill young foliage needed as food), which impose direct mortality.

Extreme weather often causes the decline or extinction of butterfly populations (Murphy *et al.*, 1990), and may be especially significant for small, isolated populations (p. 32). For example, droughts in California and Europe have caused widespread declines in butterflies; and more specifically, a late season snowstorm in Colorado extinguished a subalpine population of the lycaenid *Glaucopsyche lygdamus* by destroying inflorescences of the larval food plant, on which females oviposit (Ehrlich *et al.*, 1972). Unusually wet weather, such as that associated with El Niño events, can also cause declines.

Catastrophic weather may, of course, also affect common and widespread species, though such cases are documented only rarely. Shapiro (1991) documented the effects of a severe cold period on the Buckeye (*Junonia coenia*) in northern California. The butterfly was apparently eliminated from this part of its range, the northernmost area of its permanent residency, but much of the area was progressively recolonized during the second half of the following summer. The local population of *Erebia epipsodea* (Satyrinae) also declined substantially at that time.

Some influences of this nature have been quantified in a few population studies of particular butterflies, and a long-term stable equilibrium trend may give little cause for concern, even if there are substantial short-term variations in numbers. In a survey of weather associations with butterfly abundance in Britain from 1976 to 1986, Pollard (1988) showed a clear relationship between increased numbers and warm, dry summers. There was also some correlation between increased numbers in the *current* year and wet conditions early in the *previous* year.

Warren *et al.* (1986) studied population fluctuations of the Wood White (*Leptidea sinapis*: Pieridae) in Britain over eight years. Wet and cold weather during the flight period and time when the caterpillars were small led to fewer adults in the following year and, in formal terms, the variables were reproductive output (fecundity) and larval survival. It is extremely difficult to quantify all influences on a natural population; a large proportion of *L. sinapis* caterpillars disappeared without trace from their food plants, a loss attributed by the authors to 'predation', but virtually impossible to prove. Warren *et al.* considered that annual fluctuations depended on weather variability, and the influences of weather are indeed an all-pervading influence on butterfly numbers which can be estimated only in very general terms. It may act through influencing oviposition, predominantly constraining it (Dempster, 1983). As well as affecting numbers, weather may also influence seasonal patterns to alter the times of appearance of adults or other stages.

Causes of natural changes in butterfly numbers, commonly manifested as a 'temporary' (of unknown duration) decline, thus include weather, natural enemies, other animals (such as the availability and abundance of particular ants for Lycaenidae) and vegetational or habitat change. Food supply is affected both by weather and by the numbers of other animals needing it. Competition for food may occur as the number of species *or* the number of individuals needing it exceeds the supply of food. Some interactions of these factors are summarized in Figure 3.3.

Ehrlich and his colleagues (Ehrlich *et al.,* 1980) studied the effects of the pronounced drought in California in the mid-1970s on populations of two species of *Euphydryas* butterflies (Nymphalidae). A number of populations became extinct as a direct consequence of the effect of the drought on the availability of larval host plants on a very local scale. Droughts of similar magnitude may occur every 50–100 generations of the butterflies. This stochastic extinction of local populations may be both relatively frequent and significant in the biology of non-vagile butterflies. In these *Euphydryas* species (*E. editha* and *E. chalcedona*), migration even between populations only 50–100 metres apart is very low. In this particular case, local variability of conditions, including a division of annual and perennial food plants, led to different populations being affected in different ways by this potentially severe climatic stress. Regional persistence involves complex sequences of extinctions and recolonizations in butterfly populations of this sort, which are sometimes referred to as 'metapopulations' (p. 33).

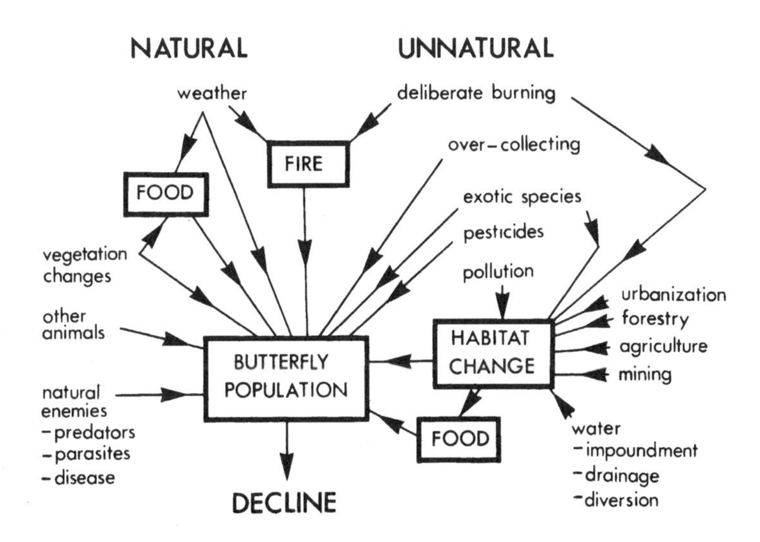

Figure 3.3 Some factors leading to decline in butterfly populations (modified from New, 1984).

Other natural calamities do occur, of course, although we shall see later that avalanches or cliff falls may help to conserve some butterflies by regenerating early stages of vegetation succession! Ash fall and lava flows from major volcanic eruptions can obliterate local communities. When Mount St Helens (Washington State, USA) erupted in 1980, the resulting ash fall covered an area of some 200 000 square kilometres to a greater or lesser extent. Its direct effects on local butterflies remain unclear. However, in Java, some caterpillars were unable to feed on vegetation covered by ash from local volcanic eruptions, and died (Gennardus, 1986).

HUMAN EFFECTS

Superimposed on these natural events is a vast array of human effects which can be regarded as 'unnatural'. Their diversity is indicated in Figure 3.3. The causes of decline and extinction of animals and plants have been discussed in many books and meetings (e.g. Fitter and Fitter, 1987); in general, three major effects can be involved:

Deliberate extirpation. This involves the deliberate hunting and/or destruction of pests, and is much more relevant to some other groups of animals than to butterflies. Few, if any, butterflies have been the concerted targets of such campaigns, although the various attempts to eliminate pest insects in environments in which butterflies occur have occasionally resulted in their unintentional decline.

Over-exploitation. Butterfly collecting is discussed later, and there are at least a few cases where it may have been instrumental in the decline or disappearance of particular species or populations. Schiotz (1987) considered that over-exploitation is commonly of such a nature that it will *not* lead to extinction (even of whales, for example, despite widely held opinions to the contrary) but may rather lead to the decline or extinction of the trade based on those species as they become rarer and more costly to obtain, thus leaving remnant populations which may survive. This condition may not apply to butterflies, as the capital investment for specialist trade tends to be small and the trade itself is somewhat opportunistic, or is replaced by individual collectors, whose zeal may not be moderated by economic considerations.

Habitat degradation and destruction. As for most other insects and other species, this is the single most important and widely recognized agent of butterfly destruction and decline. Without a suitable place to live, a butterfly is doomed, either locally or on a broader scale. These changes, itemized and discussed by Pyle *et al.* (1981), Wells *et al.* (1983), New (1984) and Ligue Suisse (1987), among others, range from deliberate and large-scale changes to terrestrial and freshwater ecosystems to highly localized and accidental efforts. In some instances well-meaning conservationists have inadvertently caused substantial harm to butterflies through influencing changes in grazing patterns, wetland drainage or woodland management, for example. A few general cases of habitat changes are noted here, and specific examples of these and others will come to the fore in later chapters. Heath (1974) regarded resource destruction as a result of habitat change as the major factor responsible for changes in range, and extinctions, of British Lepidoptera during the previous hundred years. Destruction of rainforests has undoubtedly reduced butterfly numbers in the tropics (Brown, 1976), and there are numerous records of urban expansion being associated with the disappearance of significant species. The latter is by no means a new phenomenon; there are many references from late in the last century (mainly from collectors) bemoaning this trend in several parts of the world. The 'urban sprawl' has destroyed considerable amounts of prime butterfly habitat in various countries and, as Singer and Gilbert (1978) emphasized, activities of urban humans (such as grass-cutting, clearing of vegetation, application of herbicides and insecticides, tilling, hedge-clipping, and replacement of native vegetation by exotic garden plants) can restrict both the spectrum of resources available to butterflies and their larvae and the period during which they are available. Conservation in urban environments is a specialized aspect of butterfly biology, and is looked at more closely in Chapter 8. Allied with this are such operations as the expansion of freeways, sports complexes, airports and related facilities — the general amenities needed by large human populations demanding a high quality of life. The expansion of Los Angeles International Airport, for example, has led to complex attempts to conserve a lycaenid on nearby sand dunes, the El Segundo Blue, *Euphilotes battoides allyni* (Orsak, 1982c; Mattoni, 1989). Many local butterfly colonies in Japan have become extinct because of 'devastating industrialization and recent rapid rise of the "living standard"' (Sibatani, 1989). Destruction of hedgerows and the natural field margins to increase the effective field size in Britain during the 1960s caused much conservation concern, and management of some hedges (including blackthorn) to conserve butterflies is only rarely compatible with efficient management practices for agriculture (Thomas, 1974). The drainage of fens during the mid-nineteenth century was the prime factor leading to extinction of the Large Copper butterfly in England (Duffey, 1968, and see Chapter 7); drainage of subcoastal swamps in Brazil has engendered the decline of the Fluminense Swallowtail (*Parides ascanius*), and the cultivation of marshes in North America has resulted in the rarity of the Dakota Skipper (*Hesperia dakotae*). Changes in grassland grazing regimes were implicated in the decline of various European lycaenids (Thomas, 1980; Erhardt and Thomas, 1991; and Chapter 6). Shade-loving butterflies of temperate-region woodlands have tended to decline as their habitats have been alienated by conversion to early successional open habitats or to vegetationally-alien dark commercial conifer forests. This list could be extended to an even more depressing length.

PESTICIDES

The effects of pesticide use, in particular the side-effects of widespread agricultural and forestry pesticides which have sometimes been applied with little concern for problems of drift and contamination of nearby non-target habitats, have been difficult to document. Pesticides are often cited as a major cause of the decline of natural insect populations, and their use in restricted natural habitats must be viewed with concern (Pyle *et al.*, 1981). In general, though, their effects on butterflies are likely to have been far less important than habitat destruction. Herbicides may kill food plants and nectar plants over a wide range, and their widespread use may contribute to butterfly decline. Brower (1995) predicted that they will lead to the decline of the Monarch in North America: there, herbicides are now sprayed by more than a million 'certified applicators' with the goal of killing all competing plants over tens of millions of hectares of croplands. 'Weedy' food plants of butterflies may well be reduced as a direct consequence of such extensive use. Herbicide use exceeds the combined use of all other pesticides in North America. However, as Davis (1965) noted, for general broad-spectrum pesticide applications the very insect species of greatest conservation concern are likely to be those most affected, because destruction of even a few individuals of a very rare species may decrease the breeding population by a high proportion. For butterflies in general, 'allegations exceed established fact' (Pyle, 1976). More butterflies were found to be present on 'conservation headlands' in agricultural lands in Britain than in normally sprayed fields, in a survey extending over four years (Dover *et al.*, 1990). Conservation headlands may help to protect hedges and other field boundaries against spray drift (Dover, 1991). However, the overall conservation value of headlands for butterflies may depend on allowing the growth of a wide range of larval food plants rather than using herbicides (as at present: Dover, 1991) to eliminate grasses which could be utilized by Hesperiidae and Satyrinae. The selective modification of floral balance in such areas to maintain a balance of nectar needs, larval food plants, shelter, 'correct' insolation and other resources may be important in detailed management for butterflies (Dover, 1996).

Singer and Gilbert (1978) also summarized the effects of urbanization on the predators of butterflies and emphasized the point that changes of habitat features will almost certainly affect the community of animals living there and may affect the long-evolved 'balance' between a species and its natural enemies. It is, for instance, common for people in towns to feed birds, especially in winter, and to provide nestboxes for them. These factors may help to reduce the winter mortality rate of birds, and to increase their breeding success. The result is a larger population of predators for insects. There are very few good comparative studies, but Baker (1970) found that birds were the major predators of early stages of *Pieris rapae* in a British garden while arthropods were predominant predators in more rural environments. The abundance of social wasps is sometimes increased by availability of nesting sites on buildings (although this may be countered by a greater tendency to swat them or to destroy their nests), and can lead to local increases in predation on caterpillars. Gilbert (quoted in Singer and Gilbert, 1978) observed 'intense predation' on larvae of a swallowtail (*Papilio polyxenes*) by a *Polistes* wasp in Austin, Texas, and the Preston-Mafhams (1988) suggested that these wasps may in fact be major enemies of butterflies, with their effects substantially underestimated in the past.

The effects of introduced generalist predators have only rarely been assessed in relation to butterflies. The Argentine Ant, *Linepithima* (formerly *Iridomyrmex*) *humile*, is one such species whose effects may be far-reaching. In South Africa, Henning and Henning (1992) considered it 'probably the most damaging alien introduction', because by killing off native ant species it may affect myrmecophilous Lycaenidae. It may also influence vegetation composition by destroying native ants which disperse seeds. Should the ant continue to spread, many lycaenids throughout the country could be affected (Samways, 1993).

Severe and potentially irreversible changes in butterfly numbers may thus occur as a result of many human activities. *Declines* in some species caused by habitat destruction or changes in the balance of resources needed by the species may, of course, encourage the *increase* of some other species that can immigrate or exploit the resources (such as a changed array of plants) which exist as a result of the change wrought by humans. So butterflies may indeed be common in gardens or pastures, but the species present may be quite different from those in the areas before they were changed. All too often this trend — whereby 'butterflies' do not vanish completely — may mask the decline of ecologically specialized species and populations whose demise goes unnoticed, or which are usually not noticed until the change is imminent. Many of the examples discussed later have arisen from this kind of situation and have been engendered by 'crisis management' to some populations of rare or localized taxa resulting from the immediate threat of some habitat change which would inevitably lead to their extinction.

INFLUENCES OF COLLECTING

An emotive topic noted earlier and often involved in discussions of butterfly decline is 'over-collecting'. As we shall see, legislative prevention or the prohibition of collecting is often a first step taken to 'protect' a butterfly species. Commercial harvesting or exploitation of other invertebrates — corals, sponges, edible crustaceans, some molluscs — has led to various forms of trade ban or commercial quotas (sometimes involving restrictions on size classes as well as numbers, or the introduction of 'closed' seasons) regulated by licences to control the levels at which these are taken. Butterfly collecting has long been an instructive and healthy hobby for many people, and one which has led to the detailed data base on these insects accumulated over many decades. There has been much discussion (sometimes passionate and ill-informed) over its effects. Some extremist preservationists have advocated a total ban on butterfly collecting (see p. 80), whereas many collectors claim that rational harvesting or responsible collecting (rather than the mass slaughter of insects by poisons or light-trap catches of moths which are killed rather than released) does no harm. This controversy has been around for a long time: an early insect conservation committee in Britain was initiated (Sheldon, 1925) predominantly because of concern about the activities of 'greedy' collectors. In general it is extremely hard to justify any total ban on collecting of most butterfly species on scientific grounds.

A number of 'codes' for collectors now exist, the best-known being that issued by the Joint Committee for the Conservation of British Insects (JCCBI, 1971), which has provided the model for most of the more recent codes. All advocate responsibility and restraint over the numbers of specimens taken, and responsibility for the environment and habitat of the species involved. Care is clearly needed to avoid irresponsible over-collecting because, without a sound knowledge of a butterfly's ecology, the factors rendering it vulnerable are not completely understood. Habitat fragmentation and change mean that many butterflies are now more susceptible than at any previous time in their history, as many populations are smaller, more isolated and 'stressed' to varying extents by having smaller gene pools. Several people have suggested that, at the very least, collecting can do nothing to increase the chances of survival for such populations and, at most, could tip them towards extinction. Even if, say, a collector is allowed to capture only two individuals of a rare species, the combined effect of 20–50 (or so) responsible collectors, each scrupulously heeding this 'bag limit', could result in the capture of a substantial number of individuals. Most often, however, the effects of individual collecting are much less harmful than habitat changes.

It is now acknowledged widely that simple bans on collecting play only a minor role, if any, in conservation; that broad-scale prohibitions without any scientific justification can alienate many people, and that collecting is almost always unimportant in relation to habitat loss.

As well as collecting by individuals for their private hobby purposes, or by researchers studying butterfly biology, there is a substantial amount of commercially directed collecting, and this merits some appraisal here. There are several different categories of commercial collecting, with very different clienteles and emphases. These were discussed by Collins and Morris (1985) and are:

The 'low volume/high value' category. This refers to commercial collecting of very rare or local butterfly species, with data, for sale to wealthy collectors or museums. It is extremely hard to control or monitor because the high prices available remain a substantial inducement to circumvent any 'protective' legislation or trade laws, and the small size of dead butterflies makes them easy to smuggle. Many of the species involved, as well as being extraordinarily rare, are large and showy, and over-collecting becomes a 'definite probability'. Individual specimens of some of the rarest birdwings (Papilionidae) have been offered in dealers' catalogues for as much as US$7000 a specimen. Females are sometimes much rarer than males as they are more difficult to catch, even though the natural sex ratio may be close to unity. In the past, collectors sometimes shot these high-flying insects down with dust shot, or employed local people to climb trees and wait for them. The equivalent modern tactic is to employ local collectors to wait at hill-topping sites or in forest clearings with very long-handled nets.

Many of the species offered individually for sale to collectors are not so rare, of course, but Collins and Morris (1985) listed more than 470 species of Papilionidae (more than 80% of species in the family) being offered in commercial catalogues between 1980 and 1985, 28 of which were priced at more than US$150 a specimen.

The 'high volume/low value' category. This is trade directed at the tourist or souvenir hunter. Enormous numbers of butterflies, mostly common species, are captured (predominantly in South-East Asia and South America) for this trade. The better (larger, showier) specimens in good condition may be mounted individually in clear-topped boxes for sale to tourists, but hordes of others, including many which would not be in sufficiently good condition to merit individual sale, are used in the manufacture of 'ornaments'. Many have their bodies discarded — these may be used for pig food in some places — and the wings are retained. These are sometimes laminated in plastic with painted bodies to make table cloths, placemats, ash trays, jewellery and even toilet seats, or are sometimes used unmounted to make 'butterfly pictures'.

The trade in artwork made from butterfly wings in Africa (Larsen, 1995a, 1995b) consists largely of common, pan-African species. Much of it originates in the Central African Republic, from where pictures are dispersed widely to other countries. In his analysis of one picture (a parrot of greatest dimensions 24 × 7 cm), Larsen (1995a) found 12 species of Papilionidae and Nymphalidae with a minimum use of 23 individual butterflies; parts of some of these were available for use in other pictures. The butterflies were likely to have been captured in forest clearings. Larsen estimated the annual value of the trade as around US$0.5 million for up to 20 000–30 000 pictures, so that (although it might not happen at present!) there is potential for substantial local income for the producers/collectors involved. Such articles are on sale in many parts of the world, and it is not only local butterflies which may be involved. It is indeed disconcerting to find, for example, Brazilian butterflies on sale in Kuala Lumpur as 'Souvenirs of Malaysia'.

It is virtually impossible to estimate the numbers of butterflies captured for this trade. Taiwan alone has had annual sales estimated variously at between 15 and 500 million butterflies, with a cash value somewhere in the order of US$2–30 million, and involving several hundred active dealer-collectors. The National Research Council (1983) estimated the total trade in tropical butterflies, which seemed at that time to be rising, as worth US$10–20 million a year. A higher estimate was given by Collins and Morris (1985): they suggested that an overall trade value of US$100 million a year may not be unrealistic. For Brazil, Carvalho and Mielke (1971) claimed that more than 50 million butterflies, a large proportion of which are the large, iridescent blue *Morpho* species so popular in butterfly jewellery, were killed for sale each year. Larsen (1995b) considered that this figure was 'impossibly high', but conceded the great difficulties of gaining precise information on the value of the butterfly trade.

The effects of this vast toll on natural populations are difficult to assess. On the face of it, it would seem to be one of the largest mortality factors imposed on some populations, and that it must lead to decline. However, this may not be the case, and this apparent paradox is due to the main collecting methods employed. Many commercial collectors use baits, such as carrion or rotting or fermented fruit (see p. 110) to attract and concentrate large numbers of butterflies, and some dealers appear to rely almost wholly on this field technique. Hordes of butterflies are attracted to such baits, but virtually all are males. Marshall (1982) quoted an estimate that only one in 10 000 butterflies captured in Taiwan for sale is a female, and Carvalho and Mielke (1971) gave sex ratios of various (named) *Morpho* spp. as in the range of 50:1 (male:female) to 200:1, with those of other desirable Brazilian butterflies being 10:1 (*Caligo* spp., owl butterflies) or 5–20:1 (various *Heliconius*) to 100:1 (Pieridae). Males of some species are attracted to baits because of their critical need for sodium in order to mature their reproductive systems (Arms *et al.*, 1974).

In some common species not taken by baiting, the males are brighter and more 'showy' than the females (features which make them more commercially desirable) and so are actively selected by dealers. The suggestion that this form of collecting does no harm to the populations is based on the implication that, because mainly males are taken and most males mate several times whereas females may mate only once, the breeding population is essentially unaffected. Females are left free to oviposit and most are likely to have mated, or enough males will be left to ensure that this will happen. Much more detailed information is needed on the biology of most of the species involved. As Ehrlich *et al.* (1984) have emphasized, even the definition of 'natural sex-ratio' is extraordinarily difficult. Whereas unity (equal numbers of males and females) is presumed to be the rule for most species, *Acraea encedon* (Nymphalidae) in Africa has two kinds of female, one of which produces male and female offspring and the other females only (Owen, 1970, 1971b), and the nymphaline *Hypolimnas bolina* commonly produces all female broods (Clarke *et al.*, 1975); both are very exceptional.

The 'live trade' category. In recent years, live insect exhibits — predominantly 'butterfly houses' — have proliferated in several parts of the world and have served to introduce many people to these insects at close quarters for the first time. This trend has engendered an increasing trade in mainly showy species which are not overly rare. Many institutions do not have the resources to set up permanent breeding houses for all their exhibits (though about a third are bred on site in Britain, and some are redistributed to other exhibitions) and rely on importing live material, predominantly as pupae, for exhibitions. A few species are transmitted more commonly as adults. This 'live trade' is a medium value one with low, but continuous and increasing, volume and some pressure to diversify to provide 'novelties' to attract more customers.

More than 300 species of Lepidoptera have been exhibited in Britain alone (Collins, 1987a), and the great majority of these are imported. In contrast, butterfly houses in Australia are unable to import livestock from other parts of the world. The Melbourne Zoological Gardens, for example, relies mainly on its own captive rearing programme for tropical Australian species, rather than solely on wild-caught material, in order to provide a display all year round.

Most butterfly houses seek to create tropical or 'rainforest' environments with high temperatures and high humidity, in which large showy butterflies appear natural, and both high numbers and reasonable diversity are expected by the paying visitors. Collins (1987a) noted that British butterfly houses (of which there are more than 40) attracted about four million visitors in 1986, with gross gate takings of around five million pounds. The average exhibit showed 10–40 species at a time, and from 26–75 during the season (commonly April–October). The largest butterfly houses displayed up to 70 species at once, and 150 over the season.

No species used is in any immediate danger of extinction and about half a million butterflies are utilized each year. Most were imported from the Philippines (61 species) or Malaysia (86 species), with smaller but significant numbers also from India (40 species), Taiwan (35) and the USA (32). Twenty other countries also contributed to this British market. Whereas many species of butterfly were shown in only one house, or a very few exhibits, 23 species of Lepidoptera were particularly frequent and shown in 75% of the 18 houses analysed by Collins. Two of these 23 were giant silkmoths (Saturniidae); the other 21 are listed in Table 3.1. The preponderance of large showy taxa is clearly evident, and other desirable features include longevity, ease of breeding on site and availability of regular supply. A taxonomic analysis of the butterflies listed by Collins (1987a) for British butterfly houses (Table 3.2) clearly shows the preponderance of Papilionidae, Nymphalidae and Pieridae over smaller butterflies such as Lycaenidae and Hesperiidae. The larger houses typically attempt to display 500–1000 specimens at any time.

Table 3.1 The 'favourite species' of butterflies shown in butterfly houses in Britain. (Collins, 1987a).

Species	*Common Name*	*Origin*
Papilionidae		
Graphium sarpedon	Blue Triangle	Asia
G. agamemnon	Green-spotted Triangle	Asia
Pachliopta aristolochiae	Common Rose	Asia
P. kotzebuea	—	Malaysia
Papilio demoleus	Lemon Butterfly	Asia
P. memnon	Great Mormon	Asia
P. paris	Paris Peacock	Asia
P. polymnestor	Blue Mormon	India, Sri Lanka
P. polytes	Common Mormon	Asia
P. protenor	Spangle	Asia
P. rumanzovia	—	Philippines, Indonesia
Nymphalidae		
Danaus chrysippus	African Monarch	Africa, Asia
D. gilippus	Queen	USA
D. plexippus	Monarch	North America
Idea leuconoe	Tree Nymph	Asia
Agraulis vanillae	Gulf Fritillary	South America
Dryas julia	Flambeau	South America
Heliconius charitonius	Zebra Helicon	South America
H. melpomene	Postman	Central and South America
Hypolimnas bolina	Common Eggfly	Asia-Pacific
Inachis io	Peacock	Europe

Table 3.2 Taxonomic summary of butterfly species flown in sample of 18 British butterfly houses in 1986 (Collins, 1987a).

	Total Species	*Frequency (x/18)*	*Bred on Site x/18*	*Number of Species Obtained as Adults (A), Pupae (P) or Other (C)**
Hesperiidae	3	2–4	0–3	3P
Pieridae	34	0–12	0–8	1C, 31P, 10A
Papilionidae	70	0–17	0–16	5?, 62P, 5A
Nymphalidae	142	0–16	0–16	10?, 2C, 10P, 39A
Lycaenidae	8	0–8	0–6	7P, 1A

* Some species in more than one category

The butterfly house trade seems likely to increase both in the number and the diversity of species displayed. As the public becomes familiar with the more common and easily available butterflies, 'novelties' will assume additional importance in attracting visitors. Butterfly houses are also becoming established in many other parts of the world. Ross (1996) noted that 'at least 150' exist worldwide, with a recent increase (to 24) in North America. One of the largest (indeed, its elaborate guidebook by Khoo and Chng, 1987, claims it to be 'the world's . . . largest tropical butterfly exhibition farm') is at Penang, West Malaysia, where an average 4000 butterflies representing over a hundred species fly freely in a 41 × 43 metre enclosure. More than 70 species have been reared there, and about 70 per cent of the exhibits are from bred stock.

Breeding 'in house' away from the tropics is an expensive operation, requiring the maintenance of specific food plants. Florida's Butterfly World (Emmel and Boender, 1991) is reported to cultivate some 10 000 plants, including the world's largest collection of passion-vines as hosts for *Heliconius* butterflies (Ross, 1996). Some houses are able to sell excess pupae to other exhibitors. However, the future for breeding operations in some places is doubtful. 'Butterflies in Flight', which opened in New Orleans in 1995, was reportedly issued with operating permits by the US Department of Agriculture which excluded all host plants (meaning that no butterflies could be reared), and which required that no butterfly could leave the exhibit. These regulations were imposed because of the fear of introducing any new agricultural pest or microbial disease (Ross, 1996), and it remains to be seen how the ruling may affect other butterfly houses planned for establishment in the near future.

The broad possibilities of conservation using butterfly houses are being emphasized progressively in several parts of Europe (van der Heyden, 1992).

It is very hard to assess whether commercial collecting other than that in the first category does any real harm to butterfly populations. Persistent large-scale collections associated with the second category also may have the potential to do so but, despite the well-intentioned protestations of people alarmed at the massive scale on which this occurs, the data are clearly ambivalent, and closer monitoring is needed or advisable, at least for particular local taxa. Slow-breeding species, such as some birdwings, may be more vulnerable than taxa which breed rapidly.

POLLUTION

'Pollution', in the broadest sense of contamination of the environment, has both chemical and biotic components. 'Chemical pollution' includes such things as pesticide use (see earlier) and atmospheric contamination, including acid rain, as a consequence of industrial activity. Heath (1981) showed convincingly that a number of European butterflies have declined severely in areas to the north and east of principal industrial zones of western Europe — areas almost certainly affected by atmospheric pollution borne by south-westerly winds, and these trends have continued. In Norway, the decline of the Apollo Butterfly (*Parnassius apollo* : Papilionidae) has been attributed to acid rain, a phenomenon which (although rarely documented) is likely to have drastic and far-reaching effects on many invertebrates either directly or by contaminating their food plants.

'Biotic pollution' includes the effects of exotic species — those animals or plants introduced, deliberately or by accident, from other parts of the world. Exotic animals implicated in insect decline include non-specific biological control agents (predators and parasitic wasps), birds, and mammals such as rats. The effects of these tend to be revealed most spectacularly in isolated or relatively pristine communities such as islands or less developed parts of the world. We need to remember that much of Europe, in particular, has been subject to 'civilizing influences' for so long, including exotic species introductions, that today's environments are usually not really 'natural' but already very highly modified. Such areas have long been subjected to the sorts of 'waves of extinctions' and faunal change that cause major concerns in other parts of the world which have hitherto not been subjected to such intense human pressures or change. The effects of exotic species on butterflies, other than habitat change (such as that caused by grazing animals) are difficult to document precisely. Gagné and Howarth (1982) reviewed the status of endemic Hawaiian Lepidoptera and, for a series of 27 moth species which have become extinct in historical times, implicated biological control introductions in the disappearance of 12 of these. Rabbits, by causing the loss of host plants and habitat, were suggested as important for a further six species. Predation by mice on overwintering Monarchs can be substantial (p. 163) and may be more widespread for other aggregating butterflies than is generally appreciated. Predation by birds and feral rats at the only known overwintering aggregation site of the rare Dominican danaine *Anetia briarea* led Sikes and Ivie (1995) to suggest eradicating rats as a conservation measure.

There are a few instances where butterfly species reaching new areas may have adverse effects on other native species. *Papilio demodocus*, the Orange Dog Swallowtail, is reputed to be an aggressive species which may be capable of outcompeting other butterflies. It has now spread to various islands in the Indian Ocean (Madagascar, Mauritius, Reunion) from its native Africa, moves which Collins and Morris (1985) viewed with concern. *P. demodocus* could have adverse impacts on several rare endemic species on all three islands, although this is not yet clear.

One other, often overlooked consequence of exotic introductions can be hybridization and 'swamping' of closely related native taxa. Two examples from the Lycaenidae reflect introductions to geographically isolated areas. The native *Zizina oxleyi* appears to be declining, and its range contracting, in the South Island of New Zealand as a result of hybridization with the invasive Australian *Z. labradus* (Gibbs, 1987); both these taxa are sometimes regarded as subspecies of *Z. otis*. They are certainly very closely related, but distinct. The decline of the narrowly endemic Avalon Hairstreak (*Strymon avalona*) on Santa Catalina Island, California, has been attributed to hybridization with the introduced *S. melinus*, but the evidence is insufficient to assess this fully (Wells *et al.*, 1983).

Even the redistribution of a species within its range may be unwise (p. 132). Brower *et al.* (1995) have warned of the dangers of transferring Monarch butterflies between the eastern and western populations in North America (p. 162). They advanced three arguments against transfer:

1. Transferred butterflies could carry disease into susceptible populations.
2. Transfers may confuse the understanding of numerous aspects of the species' basic biology.
3. Hypotheses purportedly being answered with these transfers (which have been made since 1992) are not answerable by the techniques.

They saw such reasons as far outweighing any currently foreseen benefits, and warned of the more general dangers of transferring organisms between distant populations. Members of any such population are in many ways 'exotic' when introduced elsewhere, even though they are clearly conspecific.

Many factors contributing to environmental change and butterfly decline are widespread. The destruction of tropical rainforests in South-East Asia, for example, seems inevitably to be leading to the extinction of many invertebrate species, and most of these events will never be specifically documented. Because temperate region butterfly faunas tend to be better known (simply because they co-exist with greater numbers of entomologists!), their historical declines have in some cases been clear, relatively well documented, and attributable to known causes. Collectively, these imply that habitat destruction or change is the paramount factor involved, that over-collecting may, very occasionally, 'tip the balance' against populations rendered vulnerable by habitat change, and that the factors which cause concern for the wellbeing of butterflies are immensely complex and often difficult to categorize precisely in any given case.

We noted earlier that habitats change as conditions change, whether by natural processes or by human interference. Larger-scale habitat changes are now heralded in the medium term by such trends as global warming and the 'greenhouse effect'. Whether or not this eventuates, and the likelihood seems significant (Pearman, 1989), the possibility must now enter into planning for longer-term conservation. The distribution of many alpine and subalpine butterflies could contract substantially, with a number becoming extinct.

The effects of climate change on butterflies are not easy to predict. Nevertheless, Dennis and Shreeve (1991) believed that many influences, especially those mediated through larval food plants, may be adverse to British butterflies, and attempted to assess the vulnerability of each species based on habitat and food plant associations. The discussion was broadened considerably by Dennis (1993), whose classification of biological and distributional features into those reflecting 'capacity' (resources to withstand changes) and 'flexibility' (ecological features predisposing ability to withstand change) provides a considered treatment of vulnerability relevant to much long-term conservation planning. The two groups of features (Table 3.3) are to some extent related, but much (60%) of the variance is specific to one group. Collectively, they reiterate that specialist species with narrow ecological and geographical ranges may be especially vulnerable. However, such changes also predispose an area for invasion or

Table 3.3 Factors which influence the vulnerability of resident British butterflies to environmental change (after Dennis, 1993).

Capacity factors

1. Range, reflecting latitudinal extent of occurrence on the UK mainland:
   1 = <25%, 2 = <50%, 3 = <75%, 4 = <100%.
2. Distribution, the proportion of 10 km squares occupied within range:
   1 = <25%, 2 = <49%, 3 = <75%, 4 = <100%.
3. Host plant (food plant) type:
   1 = monophagous, 2 = oligophagous (1 species/habitat), 3 = oligophagous (>1 species/habitat), 4 = polyphagous.
4. Food plant abundance:
   1 = substrate-dependent, 2 = patchy within habitats, 3 = ubiquitous within habitats, 4 = ubiquitous and cosmopolitan.
5. Vulnerability of major habitat seral stage occupied:
   1 = climax woodland or plagioclimatic bog, 2 = pre-climax forest, 3 = shrubs, tall forests and grasses, 4 = bare ground, short forbs and grasses.
6. Range of semi-natural habitats occupied:
   1 = <5, 2 = <9, 3 = <14, 4 = <18 (maximum number).

Flexibility factors

7. Dispersal ability:
   1 = closed populations with little evidence of movement outside colonies, 2 = colonial species with evidence of dispersal outside colonies, 3 = open population structures with evidence of frequent movements between habitat units, 4 = migrants and vagrants known to undertake long distance movements.
8. Voltinism:
   1 = biennial or univoltine, 2 = univoltine with occasional partial second broods, 3 = bivoltine, 4 = multivoltine.
9. Length of flight period for the longest brood in the year (egglaying females):
   1 = <1 month, 2 = 1–2 months, 3 = 2–3 months, 4 = >3 months.
10. Overwintering stage:
    1 = egg, 2 = larva, 3 = pupa, 4 = adult.

Note: Coding for all states: 1–4 most to least susceptible.

colonization to regions which are at present 'climatically marginal', which may be countered in part by changes in land use patterns.

Dennis (1993) stressed the broad opportunities for research offered by projected climatic change. For example, very little information is available on the tolerances of whole faunal elements, such as alpine or subalpine butterflies, and on the adaptability and tolerance limits of most individual species.

Extinction is the natural fate of many species which cannot adapt to change, and the general question of whether or not we should put effort into the conservation of species which appear to be naturally edging towards extinction, rather than into preventing many others from reaching that state, is a topic for serious moral and practical debate. As many people have stated, 'extinction is for ever', but setting rational priorities for arresting declines of butterflies — or of any other group of animals or plants — is extraordinarily difficult. Yet, in a world where logistic support, both as finance and expertise, is insufficient to support all such needs, this is obviously necessary. The frequent extinctions in industrial Europe during the nineteenth and twentieth centuries (Morris, 1986) are cause for major concern, not least because many of them are unambiguously due to humankind.

CHAPTER FOUR

AWARENESS AND CONCERN

Concern over butterfly decline is not new. Entomologists during the last century noted, and expressed dismay at, the disappearance or increasing scarcity of many species, especially those desired by collectors, and a number of early texts contain admonitions or warnings about the dangers of over-collecting. Some suggested that collectors should take only a minimum number of specimens and select these for high quality, rather than capturing all individuals of rare species. However, the key role of habitat, although the specific requirements of many butterflies have been understood in general terms for a long time, has been increasingly appreciated in recent decades, during which 'butterfly conservation' has progressively gained public and scientific support. It has developed into a sophisticated science, but one which is still in its infancy in many parts of the world. This short chapter traces some of this increasing awareness, which has come predominantly from the strong collector traditions of Europe and North America.

In the UK, the rapid decline of the Large Copper in the fens of East Anglia was documented in the 1840s, with the last specimen being reported in 1851 (Duffey, 1977; see also p. 175), and the Large Blue was thought to be imminently extinct during the 1880s (Goss, 1884; Muggleton and Benham, 1975; see also p. 171). Concern in the United States dates from the 1870s (Pyle, 1976). Pyle noted the early cases of the Xerces Blue (*Glaucopsyche xerces*) near San Francisco and of the White Mountain butterfly (*Oeneis melissa semidea*), considered at the time to be subject to collector pressure in New Hampshire. Of *G. xerces*, H. H. Behr wrote in 1875 that 'The locality where it used to be found is converted into building lots and between German children and Irish hogs no insect can exist besides louse and flea' (quoted in Pyle, 1976). In Australia,

G. Waterhouse (1897) noted that a local satyrine, Banks' Brown (*Heteronympha banksii*), 'was once plentiful near Mosman's Bay but now, owing to the progress of settlement, is rarely seen there'.

Particularly in the UK, conservation awareness was fostered progressively through the large number of collectors and several long-running entomological magazines; some of these publications are still extant, despite having started in the last century. From their early years, such magazines have served as a remarkable and detailed repository of information on the incidence of species, changes in distribution, unusual records, food plant, seasonality, life history data, and so on. Synoptic texts on butterflies were also available readily, so that newcomers to collecting or study could easily obtain fundamental information on them and learn to recognize every species of that restricted fauna. Even though conservation was not a prime theme for most of these early workers, the resulting data base has proved invaluable for later analysis, and for revealing some long-term distributional trends (Pollard and Yates, 1993). These are discussed in the next chapter, but there is an important tangential theme which merits comment here.

Much of this information has been provided by amateur hobbyists, and interest has been encouraged by the readiness with which they can identify their captures and find out about them from books and magazines. Knowledge of European and North American butterflies is now to some extent self-generating. More and more popular and scientific information is published each year, and many of the plethora of books on the 60 or so British butterflies overlap in coverage to a large extent. In addition, publishers see an established demand for more books and continue to seek them. Young people, among others, can easily enter the study of butterflies and, in an age of increasing 'professionalism' in biology, it is worth emphasizing once again that much conservation biology stems from good natural history as a foundation, and that there is an abundant need for more of this. The amateur naturalist still has a very real role to play in documenting our natural world.

In some other parts of the world, the situation tends to be rather different. There are few biologists and fewer hobbyists, all working from a more impoverished literature base, and commercial publishers do not see such a large or rapid return for handbooks and field guides. Without a market there is little chance of fostering interest by providing such books, but without the books, the interest is more difficult to generate. Collectors like to be able to identify their captures.

Compare, for example, the situation in Britain with that in Australia. Australia is now well served with butterfly books — that by Common and D. Waterhouse (1981) is definitive, and follows the impressive sequence of texts by G. Waterhouse (1932), G. Waterhouse and Lyell (1914), Rainbow (1907), Anderson and Spry (1893–94) and Olliff (1889) as progressively well-illustrated and complete works, each of which has in turn stimulated further interest in our butterfly fauna. Several local natural history and entomological societies are supported by amateur collectors. The largest of these collector groups is the Entomological Society of Victoria (with about 120 members in 1995). However, such groups in Australia do not represent more than a small fraction of the collector/enthusiast numbers so long evident in the northern hemisphere. Many tropical regions do not support significant amateur effort and the number of professional lepidopterists in the tropics is very small, although there are a number of 'Lepidoptera Societies' catering for both professional and amateur naturalists in various parts of the world. The potential for building up the kind of data base for butterflies now available in many parts of Europe is clearly very restricted in the tropics, although attempts to catalogue these diverse faunas are occurring progressively.

NATIONAL AND INTERNATIONAL EFFORTS

This is not the place to document the history of the world's numerous entomological societies, but several now take leading roles in promoting insect conservation and act as foci for co-ordinating efforts which otherwise would be very scattered. Two groups have been particularly important in the impetus which they have provided. In his informative review of the Lepidoptera conservation in the United States, Pyle (1976) traced the events leading to the formation of the Xerces Society (named after the extinct Xerces Blue butterfly) in December 1971. This was intended to be a body through which North American Lepidoptera conservation could be promoted, and now amalgamates the interest not only of North American workers but also of interested people elsewhere. More recently, the North American Butterfly Association (NABA) has helped to focus this interest further.

The Committee for the Protection of British Lepidoptera had its first meeting in 1925, and listed as its priority species for conservation seven species of butterflies which were either endangered or already extinct. Two of these are no longer of concern in the UK as there are no official moves to reintroduce (p. 132) either the Blackveined White (*Aporia crataegi*) or the Mazarine Blue (*Cyaniris semiargus*). The Large Copper (p. 175) and Large Blue (p. 171) were also of concern at that time. The only other species listed then and still of concern are the Heath Fritillary (p. 169), the Glanville Fritillary (*Melitaea cinxia*) and the Wood White (p. 160). The changing emphasis in the UK to particular species priority listings was discussed by Morris (1987) and Warren (1992). This committee was the fore-runner of the Joint Committee for the Conservation of British Insects, a broad-based group whose formation was stimulated by the Royal Entomological Society of London, in June 1968. This body has also been influential.

Conservation committees associated with Lepidoptera societies or more broadly based entomological societies now exist in various countries of Europe and America, as well as South Africa, Japan, Australia, New Zealand and others, and together provide co-ordination for activities in many parts of the world. They have helped initiate such activities as producing codes of conduct, setting priorities for habitat and species protection, rationalizing legislative protection for rare species, liaising with government bodies, providing information, organizing data, mapping distributions of species, producing publication and publicity material, organizing meetings and symposia and, sometimes, facilitating expressions of international concern. There have always been individual lepidopterists concerned about butterfly wellbeing — without their efforts, and the impetus they have given to stimulate broader interest and form the various groups mentioned above, many temperate region butterflies in particular would now be in a much more parlous state.

Today, a great diversity of people, organizations and government agencies influence conservation procedures and practices. Concerned naturalists may constitute well-informed and influential lobby groups, as evidenced in a number of cases of butterfly conservation. In the UK, Butterfly Conservation has about 10 000 members, representing an enormous spectrum of expertise and interest. Again, as Pyle (1976) noted, 'Lepidoptera Conservation in the United States has become a fully fledged movement with numerous successes and with many difficult problems to face'. This statement could be applied equally to the UK and other parts of Europe, and during the last decade or so 'fledging' has accelerated elsewhere. More problems continue to arise, but ways of solving these are also gradually being developed.

A major problem, sometimes a weakness in conservation planning, is how to maintain credibility while attending to the needs for conservation. It is very easy for emotive arguments for the defence of charismatic organisms to misrepresent, or even be substituted for, scientific or factual information. The status of a given species needs to be clarified as quantitatively and precisely as possible, and some of the approaches to this are considered in the next chapter. Objective compendia or 'state-of-the-art' works on butterfly conservation needs are being produced progressively as 'Red Data Books' and related lists. Two (Wells, *et al.*, 1983; Collins and Morris, 1985) have been produced through the auspices of the major international conservation union, the IUCN (World Conservation Union: formerly the International Union for the Conservation of Nature and Natural Resources), and a number of others have resulted from more local efforts, either as parts of broader invertebrate or faunal compendia or as butterfly volumes for particular countries. Examples are those for Spain (Gomez-Bustillo, 1978) and South Africa (Henning and Henning, 1989). The latter, to indicate the typical contents of such a regional appraisal, provided reviews of 141 species and subspecies in South Africa. Many are narrowly endemic and are not necessarily threatened directly at present, but they are thus designated as priority species for practical conservation and could easily become vulnerable. The Soviet Red Data Book, as another example, contains appraisals of nearly 70 butterflies, many of them very restricted in distribution. Red Data Books are, essentially, registers of taxa of conservation concern, and are used to help set priorities for conservation action.

These compilations represent an enormous amount of work and provide the best available documentation for the species they include. Red Data Books refer species to one of several categories to define their conservation status. It is widely recognized that such categorization raises many problems (Fitter and Fitter, 1987), including an element of subjectivity for taxa where our knowledge is poor. They are regarded as working tools, and are defined in Appendix 1. The former categories 'Endangered' and 'Vulnerable' may temporarily include taxa on the way to recovery but still not secure. 'Threatened' (in the IUCN's terminology) is a general term, and includes species in the Endangered, Vulnerable, Rare or Indeterminate groups. The category 'Insufficiently Known' is of minimal use for butterflies, as it could be applied to many species, but is of some value in indicating those which for one reason or another informed lepidopterists believe may have some conservation significance — perhaps, for example, as 'living fossils', taxa with unusually restricted distributions or peculiar or unique biological features — and for which the quality of available information may need urgent improvement because they appear likely to become threatened.

Figure 4.1 is a comparative representation of these. A species may be abundant or rare (Figure 4.1A) yet its numbers inevitably fluctuate. An unusually large decline may render it vulnerable, simply because recovery may be more difficult from a smaller breeding population, especially if any endangering process continues to operate. A larger decline will increase the difficulty and could render it endangered, or approaching the 'point of no return'. Finally, a drastic decline (Figure 4.1D) may result in extinction. In general, these categories designate a hierarchy of priorities. Butterflies included in the 'Endangered' category include species on the very edge of continued existence, but 'Vulnerable' species can move into this category with frightening rapidity. A number of butterflies *are* regarded as endangered in these formal terms: the Queen Alexandra's Birdwing (*Ornithoptera alexandrae*) of Papua New Guinea is sometimes claimed to be the world's rarest butterfly (p. 155), and *Papilio homerus*, restricted to Jamaica, was listed in the world's top twelve endangered species of all categories by the IUCN in 1988. As well as the IUCN category, reflecting imminence (or likely imminence) of loss, the degree of taxonomic isolation of a butterfly also influences the priority it may receive for conservation. Thus, the loss of the sole species of a subfamily or genus also means the loss of that higher taxon, but the loss of a species with many close relatives in the same genus may not be as important because the main evolutionary line is still present.

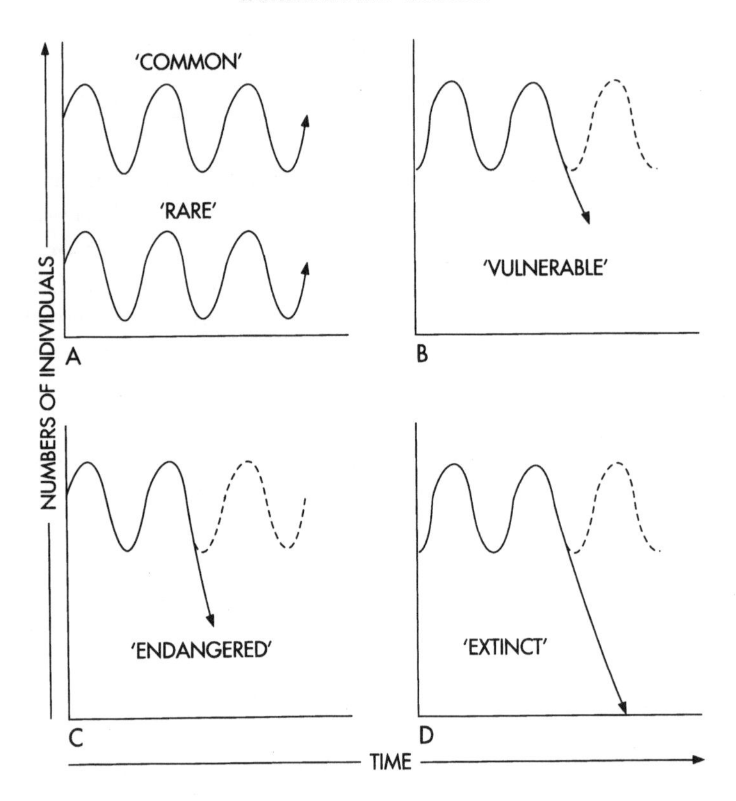

Figure 4.1 Levels of abundance and changes which may cause conservation concern (Fitter and Fitter, 1987).

Although most published butterfly status estimations use these categories, the revised IUCN categories (Appendix 1) aim to provide a more informative appraisal, based on quantitative information and estimates of extinction likelihood over a given period. They thus link with estimates of population viability analyses (PVA), a conservation technique pioneered for relatively long-lived vertebrates for which accurate information on mortality, natality and demographic profiles can be obtained in a reasonably straightforward way and for which information on decline in the wild and residual population size can be estimated. A distinction is made between the extent of field occurrence and the area of occupancy, acknowledging the fragmented distribution of many species; but because many of the parameters are difficult to determine for most butterflies, a high proportion of species will fall into the 'data deficient' category at present, even when they have been evaluated as constructively as possible.

Critical evaluation of a species' biology is fundamental to estimating its likelihood of extinction or persistence (Murphy *et al.*, 1990). For butterflies, population counts are the fundamental material of such analyses, but can only rarely be integrated to any form of reasonably comprehensive biological model incorporating knowledge of fluctuations in numbers and their causes. Many of the topics discussed in this book are included, commonly only implicitly, in such studies. Murphy *et al.* (1990) noted that, for cases where distributions are reduced but populations may remain high in remnant habitats, an emphasis on PVA must focus on the metapopulation characteristics and environmental factors which determine population persistence, and exemplified this through their study of the Bay Checkerspot.

When butterfly populations become very small as a result of habitat decline, demographic stochasticity may determine whether the population rebounds or becomes extinct. Even after nearly 30 years of research, an accurate predictive model for PVA for the Bay Checkerspot remained elusive, not least because of the immense difficulties of determining the variances associated with the parameter estimations in species which undergo substantial year-to-year changes in population size. In general, the approach is difficult to pursue constructively for most butterflies at present. A workshop held to investigate PVA for the Karner Blue butterfly (Andow *et al.*, 1994) showed considerable variations in factors influencing this subspecies in different parts of its fragmented range, but approaches similar to those used for the Bay Checkerspot led to earlier suggestions for integrating reserve design and planning for the butterfly (Cushman and Murphy, 1993).

The recent IUCN Red List of Threatened Animals (IUCN, 1994) includes 292 butterfly taxa (Table 4.1) from many parts of the world. It is widely accepted, however, that this may not really represent the magnitude of the problem — they are 'merely' the taxa we know of. A large number of other butterflies, especially in the tropics, may be at least equally threatened and may pass unheralded into oblivion as their habitats disappear. The IUCN list also indicates the minimum scale of concern; that only 2% of such a diverse group of animals can be recognized in this way may induce a feeling of complacency, but the level of recognizable 'Insufficiently Known' species additional to these

Table 4.1 Numbers of butterfly taxa included on the 1994 IUCN Red List of Threatened Animals (IUCN, 1994).

Status	*Hesperiidae*	*Pieridae*	*Papilionidae*	*Nymphalidae*	*Lycaenidae*
Extinct	—	—	—	—	—
Endangered	1	1	4	6	8
Vulnerable	4	—	23	10	15
Rare	3	1	21	75	57
Indeterminate	3	1	14	8	18
Insufficiently Known	2	—	13	3	1
Total	13	3	75	102	99

See Appendix 1 for definitions of Extinct, Endangered, Vulnerable, Rare, Indeterminate and Insufficiently Known would increase this proportion alarmingly. No 'extinct' taxa were included in the 1994 list, whereas 12 were included in the earlier (1988) version.

PROTECTIVE LEGISLATION

The IUCN categories have no legal status, but are proving to be invaluable guides for countries wishing to seek legal protection of their fauna by designation under some form of trade or restricted collection laws.

The Convention on International Trade in Endangered Species (CITES) lists several butterfly species, all Papilionidae, as being under some form of trade control. Four species [*O. alexandrae*, *P. homerus*, *P. hospiton* (Sardinia and Corsica, but see p. 80) and *P. chikae* (Philippines)] are included in Appendix I of the convention, and all other birdwings, *Bhutanitis* spp., *Teinopalpus* spp. and *Parnassius apollo*, are listed in Appendix II. Trade in Appendix I species is subject to very strict regulation, and virtually no commercial trading is permitted. Appendix II species are monitored so that the extent of trade can be regulated, and this has been the express purpose of including all birdwings there. These species may, therefore, be traded commercially, but export permits are required. Monitoring of many other butterflies in trade is desirable but, as Collins and Morris (1985) emphasized, this is extremely hard to achieve. General restrictions on trade in Lepidoptera are set down by law in various countries, such as Germany, Kenya, Malaysia, Mexico and Turkey, and many countries ban trade in particular nominated species or list these for protection. National legislation to protect local butterflies is now commonplace.

Much of this is well-considered, although some inconsistencies and anachronisms exist. Political areas with common boundaries and species sometimes differ in legislative procedures — different parts of Switzerland and Austria, and different States of the USA, for example. The limitations of simply 'listing' species for protection are being increasingly recognized, and many entomologists are strongly opposed to this practice on one or more of the following grounds.

- In the past it has been common to designate species as legally protected by prohibiting collecting or exploitation without undertaking any additional protective or research measures — that is, without providing a sound basis of ecological knowledge to clarify their status and vulnerability, or to provide for conservation or management of their habitats. This can lead to a completely false impression of the species' safety, as people tend to assume that a more active conservation programme is actually under way. Usually it is not. Such legislation, without constructive additional measures, is often regarded as a token gesture. There are cases where a species is listed as protected but its restricted habitat, so vital for its survival, continues to be destroyed because it is *not* protected.

- Collecting *per se* is not usually harmful when compared with habitat destruction (pp. 58–60). 'Listing' of a species of butterfly may serve to enhance its desirability to the small proportion of unscrupulous collectors by drawing attention to its putative 'rarity value', and so increase the commercial value of specimens. Legislation meant to protect species from collectors could create just the sort of conditions where collecting *could* cause harm to the species.
- Protective legislation is often extraordinarily difficult and expensive to police, and is often simply ignored. Wardens or other trained personnel may be needed to monitor butterfly colonies or to control quota collecting. Commonly, species listed for protection differ only very slightly from more common unprotected taxa, and the early stages may be virtually impossible to separate until they have been reared to yield adults which can be appraised by taxonomists. The practical difficulties of enforcing legislation are enhanced considerably in such cases. Whereas it *may* be possible to 'police' legislation for very distinctive taxa, it is often not practicable for others which can be differentiated only by specialists or after microscopic examination.
- It may appear at times that species selected for protective legislation have been picked out rather haphazardly or on the basis of poor or ill-advised information, rather than on the basis of rational scientific argument related to either biological peculiarity, scarcity or vulnerability. The legislation may, of course, be well-intentioned. In Queensland in 1974, birdwing butterflies (*Ornithoptera* spp.) and *Papilio ulysses* were added to the list of protected fauna under the State's *Fauna Protection Act*, seemingly because they were deemed likely to be over-exploited by collectors. These butterflies are widely distributed and not regarded as rare by entomologists familiar with the Queensland fauna, and *P. ulysses* is a popular emblem for facets of the tourist industry in tropical Australia. Indeed, it seems to be increasing in some garden environments. They were listed without consultation with the well-informed Queensland entomological community. Unfortunately this sort of thing tends, inadvertently, to alienate entomologists and collectors, which is a tragedy in the context of trying to foster responsible conservation interest. Awareness that particular species may *need* to be designated as protected is vital, but the actual designation and selection of species must be based on the best possible information.

In many parts of the world, listing of species for protection is at least reasonably rational. However, some blanket prohibitions on collecting appear to be over-zealous, and motivated in part by pressure from a rabid anti-collecting lobby. Larsen (1994) noted, correctly, that 'if their views are to be taken seriously, it is important that naturalists do not cry wolf too often', and that critical resources should not be wasted on spurious species conservation. In South Africa, all collecting of *Charaxes* spp. was banned by the Transvaal government in 1983, despite some species being common. He commented that many butterflies may be placed on Red Data Book lists because they are scarce in collections and have small apparent ranges, while on the ground they may be sufficiently abundant that 'even busloads of singleminded collectors could make no dent in the population' (Larsen, 1995a,b).

Another case is that of *Papilio hospiton*, the Corsican Swallowtail, endemic to Corsica and Sardinia and highly sought by collectors. It had been categorized as 'endangered' and given a high profile by listing on Appendix I of CITES. Surveys on Corsica (Aubert *et al.*, 1996) confirmed that the butterfly appears to be under no serious threat, and uncontrolled development on the island has been largely halted. As the only swallowtail endemic to Europe, its fate was clearly of major concern to many biologists, but *P. hospiton* is clearly not endangered in any way at present. In Sardinia, Balletto (1992) noted that collectors congregate in search of caterpillars each spring but that larvae are actively protected by local people, particularly by hotel proprietors keen to preserve the off-season income those visitors provide.

It is one thing to select or target particular rare or vulnerable taxa for protection, but quite another to designate *all* butterflies as protected. This can deter interest in a largely harmless hobby, and possibly hinder recruitment into the all-too-small pool of informed lepidopterists by deterring young people from taking an interest. Many passionate professional and dedicated amateur insect conservationists were introduced to the discipline via childhood butterfly collecting, and we can ill afford to hinder this enthusiasm.

Public awareness of protective legislation, in itself highly desirable, may result in collectors being harassed even if they are not seeking protected taxa. A further danger, or conflict, can also arise. If a large number of butterfly species are listed, perhaps uncritically, in the same category of 'protection', it is inevitable that a range of nuances of status is included — some may truly be endangered and in need of strenuous protection, while others may not be. Resources may be directed away from the conservation of the former towards the latter, and a dilution of conservation effort can result. In practical terms it may be preferable to list *only* the 'endangered' category, both to strengthen the case for their conservation and to avoid accusations of rather frivolous legislative protection for species which are really in little need of it. 'The Public' may be impressed initially at a long list of protected species for a given region, but a shorter list coupled with positive conservation action may be much better in establishing and fostering the case for conservation need. A number of species have been put on various lists because they are 'insufficiently known', and some such entries may not stand up to critical objective scrutiny. Credibility is vital for entries on any such legislative listing.

As Morris (1987) emphasized, twin (conflicting) attitudes have been apparent. One, favoured by some non-specialist conservationists, is that insects should increasingly be brought into the 'listing' of legislation, not least to advertise the fact that invertebrates are important and to increase public awareness of them. The other is that much legislation is 'conceptually weak and scientifically dubious' and does not address primary issues. There is clearly room and a need (as foreshadowed by Morris, 1976) for much closer co-operation between entomologists and other conservationists and for a more holistic approach to legislative needs, especially in contiguous areas. Self-evidently, butterflies do not recognize political boundaries!

The Lepidopterists' Society (1982) issued a statement on collecting policy which covers many of the concerns raised by those who would seek to broadly prohibit this activity. They recognized that collecting Lepidoptera is a way of introducing people to their natural environments, and has important roles in scientific information gathering, as well as being a recreational activity which — with precautions — can be pursued without detriment to the insects. Caution and restraint are needed, especially when the population may be vulnerable, and collecting methods must be responsible and relative, as must treatment and documentation of the specimens taken. Mass collecting for commercial purposes is expressly dissuaded. The major purposes of collecting were summarized as follows (Lepidopterists' Society, 1982):

- to create a reference collection for study and appreciation,
- to document regional diversity, frequency, and variability of species, and as voucher material for published records,
- to document faunal representation in environments undergoing or threatened with alteration,
- to participate in development of regional checklists and institutional reference collections,
- to complement a planned research endeavour,
- to aid in dissemination of educational information, and
- to augment understanding of taxonomic and ecological relationships for medical and economic purposes.

Nevertheless, collecting is coming to be deterred in many of the more developed countries, either by legal prohibition and regulation (p. 83) or by 'social conditioning' whereby an anti-collecting attitude becomes progressively pervasive (Stubbs, 1995). The strongly conflicting attitudes in Britain were discussed by Feltwell (1995).

Several European regions (legislation reviewed by Collins, 1987b) list for protection 'all butterflies' (Salzburg, Austria) or 'all butterflies, except *Pieris brassicae* and *P. rapae*' (Lower Austria), 'all butterflies, except white-winged pierids' (several Austrian Lander) or 'all diurnal species' (Vienna). Germany protects 'all members of the superfamily Papilionoidea as species of cultural or economic value, except *Pieris* and *Aporia crataegi*'. A Swiss ordinance applicable to the region around Laggintal prohibits the taking of all species of Lepidoptera, and prohibits the carrying of butterfly nets, with the purpose of protecting the endemic satyrine *Erebia christi*. In this case even common species cannot be collected and, as Morris (1987) noted, the emergence of such extremes is a hindrance to rational conservation progress.

In Japan, as another example, a number of species are subject to local 'protection' ordinances, the first of which was enacted in 1932 (Hama *et al.*, 1989; Sibatani, 1989). There are important exceptions to my implication of 'mere listing'. The US *Endangered Species Act* (1973) aims to conserve species through implementing programmes which are designed to aid recovery and survival of 'endangered' species (those in danger of extinction) and 'threatened' species (those likely to become endangered in the forseeable future overall or a significant part of their range) and the US Federal Register of Endangered and Threatened Wildlife and Plants is continually being updated. Seventy-one butterflies were listed for review in the 1989 Register, and three categories of listing are recognized. Category 1 comprises taxa for which substantial information is available to support the listing but for which detailed rules have not yet been issued. Category 2 are those which are not formally listed but for which proposals for listing are considered likely to be appropriate on the basis of information available. Further biological studies are needed to clarify the status of these taxa, and the major aim of listing is to encourage awareness of this need. Category 3 are those taxa which are considered to be extinct, those which are now taxonomically not as discrete as previously supposed, and those which are more widespread and less vulnerable than previously thought.

The 1989 notice for Review contains only one butterfly, the Uncompahgre Fritillary (*Boloria acrocnema*, pp. 167–8) under Category 1. Eleven butterflies are listed under the Act but listings ceased over much of the 1980s, and the considerable backlog of nomination is yet to be appraised.

Proposals for listing under the US legislation include the need, under 'Prime Objective', for a suggested recovery plan, with detailed specific recommendations on management to conserve a target species. 'Threats' also need to be documented and the 'request for review' solicits information on recommendation of 'critical habitat'. A 'critical habitat' is also established, so that the 'endangered' status afforded to species here is one perceived as a context for positive practical action. The US Act has attracted considerable comment in recent years, with some doubts over its effective future after a period during which little progress was made (Opler, 1991, 1995; Wilcove *et al.*, 1993).

Australia's *Endangered Species Protection Act*, operative since 1993, also carries the obligation to investigate the conservation status of listed taxa and prepare a management or recovery plan as needed. To date (October 1996) no butterflies have been listed — not through lack of need (New and Yen, 1995), but because a major review of invertebrate conservation in Australia is in train.

Another example is the *Flora and Fauna Guarantee Act* enacted in 1988 in Victoria, Australia. This provides for inclusion of invertebrate animals and non-vascular plants in the government's undertaking that 'no native animal or plant will be permitted to become extinct in the State'. A species or community may be nominated by any individual or organization and become the subject of an 'interim conservation order' which may restrain development or threatening processes to a habitat, or restrict the taking of a species, pending clarification of the status and need for conservation by the relevant State Government department. Several butterflies are listed under the Act, and some are now actively being appraised to clarify their status and prepare management plans. The Eltham Copper (p. 182) is one such species. A scientific advisory committee to the Minister includes biologists of many persuasions.

These legislations exemplify an important transition from just listing species for protection to more practical adjuncts to conservation, linked with research and practical management programmes which are likely to enjoy the confidence of both biologists and the public. The transition is inevitably both difficult and expensive — gaining the knowledge needed merely to confirm the rare or vulnerable status of a butterfly may itself be time-consuming and necessitate original research. The need to understand and arrest a decline is likely to add substantially to the scope of this, and the need for detailed knowledge before acknowledging a need for conservation can be counter-productive. A fine balance exists between the need for information to justify legislative listing and protection, and the danger of deterring people from nominating genuinely worthy species because of the top-heavy bureaucracy militating against this and making any case too logistically difficult to prosecute effectively within a reasonable time.

The role of protective legislation is, essentially, two-fold. First, it must increase public and political awareness that particular taxa really do need protection, in the context of the wellbeing of their habitat and the resources on which they depend. Second, it must facilitate, perhaps by relieving immediate pressure of endangering processes or providing an interim conservation order and site protection, the gathering of information needed to plan and prosecute conservation of those taxa.

Legislation without the second proviso can be no more that a contentious stop-gap, even if sincerely intended to protect extremely worthy species. It can even be counter-productive: Vasainanen and Somerma (1985) noted that the legislative protection of the Clouded Apollo (*Parnassius mnemosyne*) in Finland has led to a distinct *decrease* in information, 'a protected species appearing not to interest most lepidopterists'. Many amateurs, in particular, are deterred by having to apply for permits through complex bureaucratic channels.

A major task being undertaken by the various specialist groups of the IUCN's Species Survival Commission is to prepare 'Action Plans' for the conservation of the animals and plants which they consider. A preliminary Action Plan for the swallowtail butterflies (New and Collins, 1991) relied heavily on data included in the comprehensive appraisal by Collins and Morris (1995), but has helped to stimulate surveys and appraisal of some notable species. Action Plans have no legal status, but they aim to clarify the priorities for conservation within the groups of taxa concerned, identify species and habitats which should be at the forefront of conservation effort, and suggest how this might be pursued practicably. Action Plans are disseminated widely and have the potential to provide local decision-makers with rational and well-organized foci for local conservation activities which are seen to be of international significance. All levels of need may be expressed — ranging from need for documentation of species considered significant (for any reason) and the need to survey particular habitats to determine what species are present, to the urgent need for habitat reservation or species management and recommendations for the effective ranching of particular butterflies (p. 129) and/or their effective legislative protection. Several butterflies are also candidates for global 'Heritage Species' listings, again in the expectation of promoting their effective conservation on an international scale.

The methods and procedures involved in obtaining the sound knowledge of butterfly biology needed as a basis for their rational conservation are considered in the following chapter.

CHAPTER FIVE

STUDYING BUTTERFLIES FOR CONSERVATION

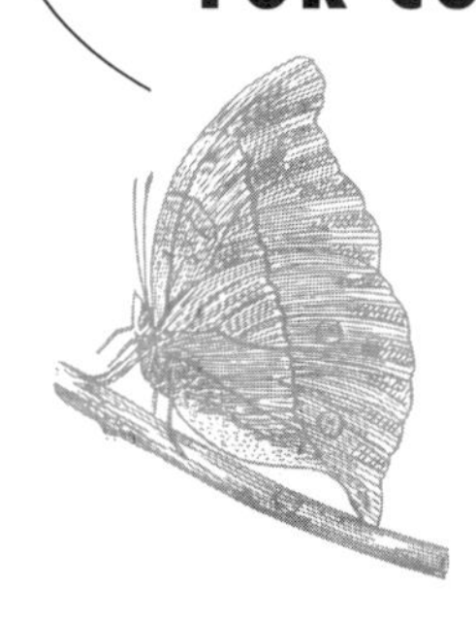

In order to assess the needs for, and plan the conservation of, butterfly species, two fundamental questions need to be answered:

1. Is the butterfly endangered, vulnerable, threatened or likely to become so as a result of any current or planned changes to its habitat and environment?
2. Can this situation be alleviated or prevented by determining and then removing or stopping the endangering processes?

In practice, answers to these and related questions depend on estimates of the abundance and distribution of the butterfly and how it fits into its environment, what foods and other resources it needs, how it behaves, and how its biological peculiarities may be satisfied. We have seen that the ecological characteristics of any particular butterfly may be complex, and it is rarely, if ever, wise to extrapolate information from related species or different populations except in rather general terms. However, the need for sound — usually quantitative — information as a basis for conservation decisions, whether concerning a single species or a diverse community of butterflies, is leading to increasingly sound methods for studying and counting butterflies in the field. This chapter summarizes some of this progress.

A SCHEME FOR PRACTICAL CONSERVATION MANAGEMENT

The sequence of steps in evaluating the conservation needs of a butterfly species consists of determining the status of the species, the threats to it, and the management needed to alleviate these, together with provision for monitoring the progress of management. The steps shown in Figure 5.1 incorporate most of the

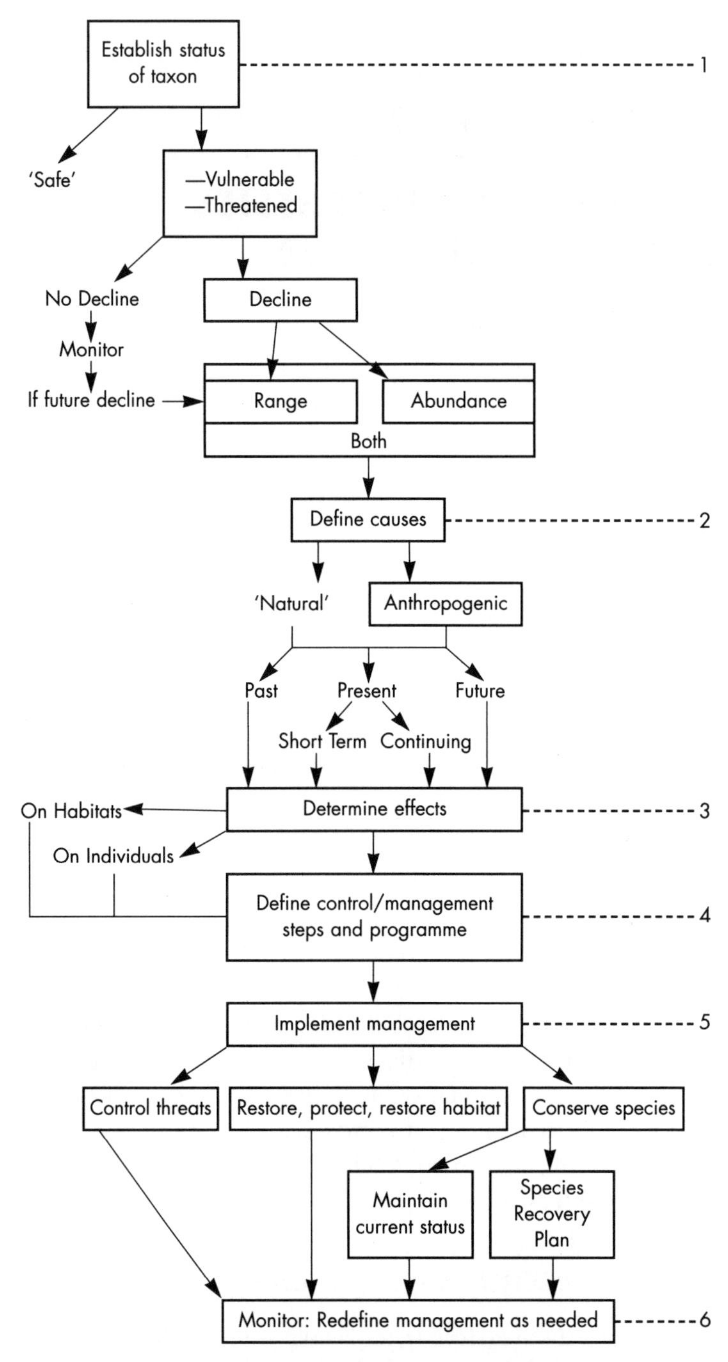

Figure 5.1 Management scheme for practical conservation of a species (after New, 1995). 1–6 are the sequential steps needed.

studies discussed in this book, though these cases have reached different stages in this overall scheme. Status evaluation is the key need in assessing conservation needs, and must depend on sound ecological data and methodology to determine distribution, population sizes and trends in abundance, together with the major reasons for any decline.

Except in obvious 'crisis management' cases (such as the imminent destruction of a key or favoured site or habitat patch by 'development', so that rapid and decisive action might be needed), the definition and implementation of management is needed to control threats, safeguard the habitat and conserve the species directly. This may be achieved either by maintaining its current status and curtailing further decline, or by a more aggressive 'recovery plan' to build up numbers and increase its distribution — perhaps by translocation or reintroduction to additional prepared secure field sites (p. 132). Provision for monitoring all such management is critical, not least to determine the reasons for success or failure so that they can be taken into account in later attempts at conservation for the same or related species. Programmes of this nature require a long-term commitment (as at least several years of monitoring may be needed) and the capability to refine the management if monitoring reveals additional biological subtleties.

The translation of these generalities into a species 'action plan' or similar document entails drawing on all available information on the species and focusing additional research on filling the gaps in knowledge. A preliminary plan is, in essence, a working document which may be refined considerably as new data accumulate, and is invaluable as a basis for discussion by all concerned parties. Thus, for the Eltham Copper in Australia (p. 182), a preliminary evaluation of status and conservation needs by Crosby (1987) was followed rapidly by more comprehensive draft and more enduring management plans (Vaughan, 1987, 1988). Each of the preliminary documents was important in focusing the conservation needs, eliciting political support, and assuring legal recognition of the butterfly (p. 183), and more species are progressively being appraised in this way (New and Yen, 1995).

In the UK a current project (initiated in 1995) aims to compile strategic action plans for 25 threatened species. These identify clear actions and responsibilities for achieving their full recovery in the UK over the next 5–10 years. Wide consultation occurred in the preparation of these documents, and there is provision for annual monitoring and periodic reviewal of the action plans. Each gives a clear statement of status and priorities, a series of broad objectives, summarizes existing relevant biological knowledge and threats, details conservation to date, and sets out a series of prioritized actions and work needed. These are divided into a number of categories (compare with Arnold's (1983) management scheme: p. 199), as follows:

- policy and legislative,
- site safeguard and acquisition,
- land management,
- species management, protection and licensing,
- advisory,
- international,
- future research, survey and monitoring,
- communications and publicity, and
- review.

The broad objectives of each plan reflect the needs of each species and the stage of present conservation activity. Thus, for *Maculinea arion*, a conservation target for many years (p. 171) they are as follows (Barnett and Warren, 1995a):

- maintain and enhance existing colonies,
- create a viable network of Large Blue colonies in the UK by establishing the butterfly in a minimum of 10 sites, based on its former distribution (detailed in the action plan),
- establish an upward trend in numbers of colonies and range, and
- encourage the conservation of all Large Blue butterflies throughout their world range.

For the Northern Brown Argus (*Aricia artaxerces*), a relatively recent target species in the UK, the objectives are somewhat different, with emphasis on the need for more knowledge (Ravenscroft and Warren, 1996):

- maintain the present range,
- halt the decline in northern England and maintain and enhance all known populations,
- increase the biological knowledge of the species, especially its conservation status in Scotland,
- encourage the restoration of suitable habitats in former range, and
- in the long term, restore the 1970 range (reflecting a substantial documented recent decline in distribution).

STATUS EVALUATION

Butterfly collectors can generally comment reasonably accurately on the status of most (temperate regions) or some (tropical regions) butterfly species in areas where they have collected regularly. They may be able to state with considerable confidence that 'species X' is common and widespread while 'species Y' is not. However, even for rare and local species, the detailed distribution and numbers may not be clearly known, except perhaps in parts of the northern hemisphere. Collectors tend to return year after year to well-known 'traditional' localities for particular species, and much of our knowledge of butterfly biology has come from the accumulated efforts and records of amateurs. The sorts of records gathered over a century or more of collector intelligence can now be effectively sorted and co-ordinated to show changing trends of abundance and distribution in parts of Europe and North America, but this state of awareness is still far off in many other parts of the world. With the urgent need to clarify the status of many butterflies, however, both on their own account and as adjuncts to broader scale habitat evaluation and comparison, sound data are now needed in a greater range of contexts than ever before.

Determining any changes in abundance of very rare species is virtually impossible, and even determining their presence can be very uncertain. For example, in surveys for the Small Ant-Blue (*Acrodipsas myrmecophila*) at Mount Piper in Victoria (p. 180, Britton *et al.*, 1995), only about five individuals were seen in three full seasons inspections, and on most visits no butterflies were seen.

The study of very rare species thus poses substantial problems (Main, 1982), including the following:

- They are very difficult to find in large or moderately large numbers.
- They are consequently costly to study, and expensive data may still be poor and not scientifically or politically persuasive,
- It may not be possible (or wise) to capture any (many) for study, so that statistically adequate data on population sizes are usually not available. Such species therefore command a conceptual shift to 'presence/absence' information.
- It is sometimes uncertain whether populations will persist long enough to provide the information needed to plan constructive management.
- Even when the biology is known, the population may be at the mercy of chance environmental events.

Defining the components of status and decline, and our capability to assess these and respond effectively when necessary, are the central challenge of species-level conservation. The development of protocols for this is advancing steadily, and these need to satisfy two major needs (New, 1996):

1. to have sufficient generality in execution and interpretation that they can be employed by non-specialists, sometimes at short notice in response to crisis needs, and
2. to incorporate sufficient flexibility to be applied to different taxa and habitats, and different time scales, without losing integrity.

Although such regimes may be employable in long-term planned studies, they may also be needed for rapid evaluation; for example as a result of short-term moratoriums on the development of important sites to allow an evaluation of their significance as butterfly habitats. Episodes such as those at San Bruno Mountain for the Mission Blue (involving *Plebejus icariodes missionensis*), Los Angeles International Airport (the El Segundo Blue, *Euphilotes bernardino allyni*) and Eltham, Victoria (the Eltham Copper, p. 182) (Arnold, 1983; Mattoni, 1992; Vaughan, 1988) are practical tests of our capability to respond in sound scientific terms and may become more frequent in the future.

Within the broad framework of butterfly conservation, status evaluation may be needed for any of a variety of purposes, and the degree of detail needed may differ between these, or between individual cases. Some examples (after New, 1996) are:

- to demonstrate a conservation need, in response to a particular threat such as development or destruction of a habitat patch,
- to designate priorities for allocating funds, or to make a case that a particular taxon should receive preferential treatment,
- to rank a series of habitat patches or sites in terms of their 'notable' species,
- to provide basic information on notable taxa as baseline information for pre-emptive use in countering anticipated threats,
- to provide a foundation for optimal management or recovery programmes,
- to provide foundation data for long-term studies of geographical change,
- to monitor the effects of long-term species, site or habitat management, and to refine such management as necessary in response to changes in status, or
- to monitor the fate of introductions or translocations.

Each of these examples depends on some level of determining some or all of the following components of 'status', even though they may not need always to be formalized for any 'legal' ranking and priority (p. 178):

- taxonomic integrity (p. 13),
- rarity, incorporating abundance and distribution. (It is worth reiterating that some reasonably precise interpretation is needed in each case, because of the numerous ways in which the term is used (p. 35). Species may be rare but stable.),
- decline, and
- vulnerability.

Evidence of historical or more recent decline may reflect solely historical events and processes, or it may reflect continuing ones (in which case the species may be increasingly vulnerable). A past decline, even if substantial, does not necessarily cause vulnerability, although it may well do so; for example, a reduction in habitat may render sites increasingly susceptible to invasion by harmful exotic species. The scale of investigation is also an important consideration. Local events and cases can assume massive local importance, and a high proportion of the practical cases of butterfly conservation have focused on local to national levels, rather than on a species throughout its entire international distribution. Local threats to a more widespread species, however devastating, may be relatively unimportant to the species on a larger scale than either (1) threats to the sole remaining population(s) of a butterfly or (2) more widespread threats likely to have an influence over much or all of a species' range.

CONTEXTS

The practical needs in three major conservation contexts involving butterflies can be recapitulated as:

1. Assessing the size and distribution of a given butterfly population, and whether its resource needs can be satisfied in perpetuity.
2. Assessing the numbers and relative abundance of the butterflies of a given area or habitat, usually either a reserve or an area being considered for possible reserve status, as a 'mirror' of the complexity of the community present. In surveys in which many butterfly species are present, changes in abundance overall may be reflected more reliably in changes in the more frequent species. For example, Swengel's (1996) study of prairie butterflies in the central USA yielded 90 species, but 34 species (each observed more than 99 times and at more than five sites) comprised more than 93% of the total (= 80 906 individuals) and were defined as the 'study species'.
3. Comparison of the butterfly faunas of different areas. These may, for example, be different vegetation types in a country or major reserve such as a National Park as a basis for management practices; of 'before and after' scenarios (such as *Eucalyptus* woodland and introduced *Pinus radiata* plantations in Australia) to determine the influences of planned or actual environmental change and the wisdom or effects of extending these to larger areas or greater intensity; or of a series of similar vegetation patches in order to help rank these in some way, either as butterfly habitats or as relative priorities for reserves or change.

Many quantitative studies, including some of those referred to in Chapter 2, have emphasized the acquisition of a fundamental understanding of butterfly biology — and many, although not originally intended as conservation studies, have become highly significant in this context as their target species have become rarer or of conservation concern. 'Basic' research in butterfly biology, as in so many other areas of biology, has the demonstrated potential to provide answers to practical questions at a later stage. Many aspects of butterfly behaviour, for example, are now recognized as important in considering conservation and the maintenance of utilizable habitat. Hill-topping sites, or their equivalent, may be needed to ensure mating, and the tolerable size and degree of fragmentation of habitat can only be determined once we know the usual population structure, how readily a butterfly disperses and how the female seeks oviposition sites.

PRACTICAL STUDY OF BEHAVIOUR

The study of butterfly behaviour is thus an integral part of determining how a butterfly interacts with its environment and the resources it may need. The main practical problem with studying the behaviour of active adult butterflies is that they *do* fly, generally fast and where the observer is not watching or anticipating, and many interactions with other individuals or other organisms take place very quickly. It is usually very difficult for a single butterfly-watcher (or even a team) to record comprehensively everything that a butterfly does; in general, relatively long durations or sequences of observations are needed to counter a bias towards a particular kind of behaviour or activity which may occur only at a particular time of day. When following the behaviour of a particular individual it can be very useful to mark it conspicuously and uniquely and, as various workers have pointed out, a grid or line of numbered posts or markers on a particular site can make plotting butterfly movements considerably easier. Dover (1989), while studying the behaviour of butterflies on field margins, divided the margins into six-metre plots, using canes bearing a small aluminium panel with a number at the top.

Note-taking, as well as being time consuming, may distract the observer's attention from the butterfly — inevitably at a critical moment! Tape recorders can be a very worthwhile substitute. Dover (1989) advocated using a small hand-held, battery-operated recorder which can record up to two hours of observations on a single tape. 'Slide' pause control is more convenient than 'push-button' controls when transcribing rapid or frequent behaviour (such as feeding, when the insect moves rapidly from flower to flower), and a built-in microphone ensures that the whole machine can be held and operated in one hand. A battery-powered digital stopwatch is ideal for timing observations. Prefatory comments (place, site details, date, time, weather, species, etc.) can easily be made, and a detailed description of the behaviour can be facilitated by using a series of 'codes'. These can be designed to suit the needs of the study, and the operator must be thoroughly familiar with them before field use, to minimize the risk of ambiguity. Dover gave examples to show how such a scheme can work. A number of activities were each denoted by a capital letter, with 'modifiers' (descriptive terms) in lower-case letters used to qualify these further. A number of punctuation codes are also needed. This shorthand was sufficient to encode all activities observed to a limit of 0.5–1 second/activity, which allows for

accurately analysing behaviour sequences and evaluating activity for comparison between species and habitats. Dover's codes may be summarized as follows:
Activity. B — basking; O — ovipositing; M — mating; N — feeding; S — settled with wings open (i.e. not B); W — walking; F — flight; I — interacting; T — terminate.
Modifiers. (for activities as indicated):

- B-W: oh — on hedge shrubs; h — on hedgebank flora, soil; c — on headland flora, soil. (These, of course, reflect Dover's study habitat: they can be amended for any vegetation type, perhaps with codes for habitat subdivisions and preferred or dominant floral species, and so on. Using a small number of categories is, in practice, far better than trying to remember many.)
- F: h — along hedgerow; c — over headland.
- I: superscript for the number of interacting individuals; can also use conventional signs for male and female.
- T: a — astray or lost; f — flew out of study area into field; g — flew through gap in hedge; h — flew over hedge; l — left study area; o — stopped recording; u — unable to identify following interaction.

Punctuation marks. E (x) — entry statement, in plot 'x'; ( ) — duration in seconds inserted; ';' — separates different activities.

Using this scheme for a male *Heteronympha merope* in open woodland near Melbourne, I transcribed 'E (5): Fgr (54); S (16); B (130 inc N [dandelion] 17); Fgr (8), I 2 ♂ (18); Tu'— which was later expanded to 'Butterfly entered the area in plot 5 and flew over grasses (= gr) for 54 seconds, settled for 16 seconds and basked for 130 seconds (including feeding for 17 seconds on a dandelion). It then flew for eight seconds and interacted with two other males for 18 seconds, after which it could no longer be recognized'. This individual was not marked and was not distinguishable after a frenzied encounter with three other males. Dover (1989) gave a more extensive example in his paper.

More sophisticated methods for recording butterfly behaviour are of course available, but are not vital in order to be able to contribute very worthwhile and relevant observations. In their study of oviposition behaviour by females of Australian *Eurema* species (Pieridae), Mackay and Jones (1989) used a small portable computer as a data logger. Each time the butterfly under observation changed its behaviour, a key corresponding to the new behaviour was pressed; the computer stored the identity of that key and the elapsed time in milliseconds. Their computer was also programmed to make an audible beep every 30 seconds, at which time a coloured flag was placed along the flight path of the butterfly, so that distances and direction travelled over a 30 minute period could be related directly to habitat features.

ESTIMATION OF POPULATION SIZE

Butterfly population size, vulnerability (as reflected in trends in size) and structure together constitute perhaps the most urgent information required on any possible conservation target. At its most fundamental, the kind of population structure essentially determines a butterfly species' ecological flexibility, with closed populations being especially vulnerable. Even limited gene interchange between partially isolated populations may counter habitat patchiness, and the extent of this can be detected only by proving whether or not individual breeding butterflies fly between putatively discrete populations, or units of a broader metapopulation. Edith's Checkerspot (*Euphydryas editha*) at Jasper Ridge, California, has been studied since 1960 (Ehrlich and Murphy, 1987). Three separate demographic units are present. These are populations which are sufficiently isolated to have independent population dynamics, but which still interact. Another, *Erebia*, *E. epipsodea*, is distributed over areas which may be very large (up to hundreds of square kilometres) which contain ecologically unsuitable areas. The butterflies have a very high level of individual movement, and the population is effectively continuous over this large range. Unsuitable areas do not constitute substantial barriers to their dispersal (Brussard and Ehrlich, 1970a, 1970b).

When studying butterfly populations it is necessary to maximize limited logistics in order to obtain as much information as possible in a reasonable time. Studies tend to be of strictly limited duration; that of Ehrlich and his colleagues on *E. editha* is highly unusual, and most studies do not extend beyond (at most) the duration of a usual Ph.D. project (three to four years). Butterfly biology is an area in which professional scientists and amateurs have traditionally co-operated extensively. Often, amateurs have initiated much of the need for conservation awareness, and much of the important data gathering and interpretation. It is vital to sustain the symbiosis and co-operation between these groups, as efficient assessment of many butterfly populations and habitats can only be enhanced in this way. Sadly, as 'natural history' becomes supplanted by more formal 'ecology' in educational institutions, these traditional liaisons are beginning to break down. For any sound appraisal of a butterfly, at least two or three flight seasons are needed for study, but the urgency of much conservation-motivated work commonly does not permit even this, and only one season (however atypical it may be) may be available — in some cases, even less.

'Marking' of butterflies has played a major role in estimating population sizes and the longevity and dispersal of individual insects. Butterflies can be marked so that they can be recognized, either as individuals or as having been captured on a particular date, then released into the wild and later recaptured. This technique can provide data which is otherwise very difficult to obtain and, if the marked individuals are of known ages, on survival rates as well. It is, of course, important that markings should not affect the normal longevity or behaviour of the insects. They should not hamper movement or behaviour, be chemically toxic (such as some paint solvents, possibly more significant for freshly emerged butterflies than for mature field-collected ones) or interfere with their natural camouflage or other putative defences against natural enemies. For butterflies, it is often useful to have marks which are conspicuous and can be detected without direct capture, although this itself may mean that these individuals are subsequently over-sampled in relation to their frequency in the population. The mark also needs to persist for the life of the insect or, at least, for the duration of its part in the study. Additionally, the marks must be applicable, and the insect handled without causing harm — for small delicate butterflies such as Lycaenidae this may be extraordinarily difficult and, as Murphy (1989) has emphasized, application of such techniques to these could be unwise as resulting mortality may be much higher than suspected. Murphy's sobering question 'Are we studying our rarest butterflies to death?' merits serious appraisal, in view of the regularity with which these mark-release-recapture (MRR) techniques are used to assess population sizes of rare species.

MRR methods are not suitable even for all large butterflies because some are strong fliers and difficult to catch, and some range over very wide areas so that populations cannot be defined easily (Pollard and Yates, 1993).

Some of the practical problems of marking butterflies have been addressed by Morton (1982, 1984), based on a study of marking the Marbled White (*Melanargia galathea*: Satyrinae) in different ways with fine-tipped marker pens. Each butterfly (25 males, 25 females in each treatment) was given a unique number on the underside of the wings, and the effects of colour (red, green or black), size of mark (small: 1–4 by 1–2 millimetres or large: 10–12 by 1–5 millimetres) and subsequent disturbance by handling on recapture were assessed. Marked individuals were allowed to mix in the natural population for two days and were then recaptured. In this species, the addition of brightly coloured marks to the cryptic underside seemed to make little difference to the likelihood of an individual's chance of being recaptured, and the size of the marks did not affect recapture frequencies significantly. However, the effect of handling appeared to be more serious, and significantly fewer 'handled' specimens were recaptured than 'not handled' ones.

In the Uncompahgre Fritillary (*Boloria acrocnema*) Gall (1984b) found that individual butterflies needed a substantial recovery period after being marked. Individuals were sometimes found resting at the release point one to two hours after being marked, and Gall attributed this (which did not appear to increase mortality) directly to handling at the time of initial marking. On later occasions, details of recaptured individuals could be obtained while the butterfly was still in the net, and not actually handled. This 'marking trauma' is restricted largely to the day of marking, but Gall suggested that it could be of wider relevance in

Lepidoptera studies. Various influences might be present here — the size, and flight ability of the species, the clumsiness or care of the individual worker, and the time between capture and release. The latter may be particularly important for sedentary alpine species. *B. acrocnema*, for example, depends heavily on exposure to sunlight (Gall, 1984a), and deprivation through being retained in a container or cage for some time could interrupt its normal daily activity regime.

Marking studies are logistically intensive, because of the time needed to mark large numbers of individuals, and large conspicuous marks may be both easier to apply and easier to detect later. Marking techniques used for assessing population size assume that the probability of recapturing marked and unmarked individuals in relation to their relative abundance is equal. Butterflies must almost always be captured in order to mark them, and the effects of physical capture and marking may be confused. Occasionally, it may be feasible to mark without capture, as Singer and Wedlake (1981) showed by marking the Green Triangle (*Graphium sarpedon*: Papilionidae) congregating at pools by a river. Similar methods can be used for some Lycaenidae which roost gregariously and may be quite motionless in the evenings, resting on grass stems or on other vegetation (Morton, 1982).

Several workers have found that marking with a felt-tip or other marker pen is unsatisfactory for small or dark butterflies because scales may become detached onto the pen, or the mark itself may be difficult to see. These species can sometimes be marked with bright nail polish, though this sometimes flakes off. Quick-drying cellulose paint was used in a number of early studies. For the lycaenid *Heodes vigaureae*, Douwes (1970) removed scales from marks on the upperside of the wing, and insects marked in this way were easily detectable from distances of one to two metres so that it was unnecessary to recapture them. In some studies it may be desirable to make some estimate of the condition of each butterfly captured, so that evidence of increasing wear or ageing can be gained over a study. The number of categories used should be defined carefully. One series of groupings used by Watt *et al.* (1977) for studies on *Colias* (Pieridae) is as follows, based on degree on wing-wear:

- very fresh — recently emerged, wings still soft and shining,
- fresh — wings dry and hard, but no visible wear,
- slightly worn — noticeable wear of scales from wings or body,
- worn — wings with frayed edges or some tearing,
- very worn — wings with extensive scale wear, fraying and tearing.

Some of the practical contexts in which MRR techniques have been applied are described here.

Studies of migration patterns of well-known long distance migrant butterflies. Either a group of butterflies in one place, or on one occasion, can be mass-marked or, more commonly, individual coded marks are applied so that each individual can be recognized.

Urquhart's pioneering studies on *D. plexippus* migrations in North America, which led to the eventual discovery of its overwintering sites in Mexico (pp. 161–5) necessitated recognition of individual butterflies over an area vastly greater than could be policed by a single observer, and also that observers hundreds of kilometres from the butterfly release point could report details of that particular butterfly. Urquhart affixed numbered adhesive paper tags, also bearing the address to which recovery information should be sent, having previously removed the scales from a short length along the anterior edge of the forewing to increase adhesion. A similar scheme was adopted by the Australian Museum, Sydney, in a study of Australian *D. plexippus* (Smithers, 1977), in which labels were glued to the underside of the hindwing. Both these studies relied on publicity to generate awareness by a network of observers requested to return any marked butterflies, or details of labels, to the host institution. Nielsen (1961) used a rubber stamp to consecutively number individuals of the migratory pierid *Ascia monuste*.

Direct labelling as used in the monarch studies is possible only for large, strong butterflies, and may be warranted only when there is need to disseminate written information rather than just a coding interpretable by the experimenter.

Local population movements. More commonly, it is smaller butterflies on a more local scale which are being studied. Felt-tip pens or nail polish type paints are now widely used for the basis of individual recognition to detect local movement or persistence (longevity). Numbering codes have almost always been based on different colours or on marks on different positions on the wing to consistently represent different numbers. A '1–2–4–7' system developed by Ehrlich and Davidson (1960) was modified by Brussard (1971), so that well over 1000 individuals can be recognized if marked with only one colour (Figure 5.2). Many of these markings are near the tips of the wings, which may become ragged as the insect ages, and a system of different coloured spots on the hindwing discal area underside can, in part, obviate errors due to this, as well as reduce the need for capture in order to inspect all four wings.

Gall (1985) advised strongly against marking codes and recommended writing numbers directly onto the butterfly with a felt-tip pen, to obviate errors in reading codes. Every butterfly would then have a direct unique mark and be individually recognizable, but closer inspection may be needed to detect these later.

In estimates of population size. The need here is commonly not for individual recognition, but for a proportion of the population to be marked in a similar way so that they may be collectively recognized as a distinctive cohort of individuals present on a given occasion. They are released and assumed to mix freely in the population, so that a random collection of individuals at a later date will include a mixture of marked and unmarked butterflies. All individuals captured on this second occasion can then be marked identically — with a mark different from that used on the first occasion — and the process repeated to

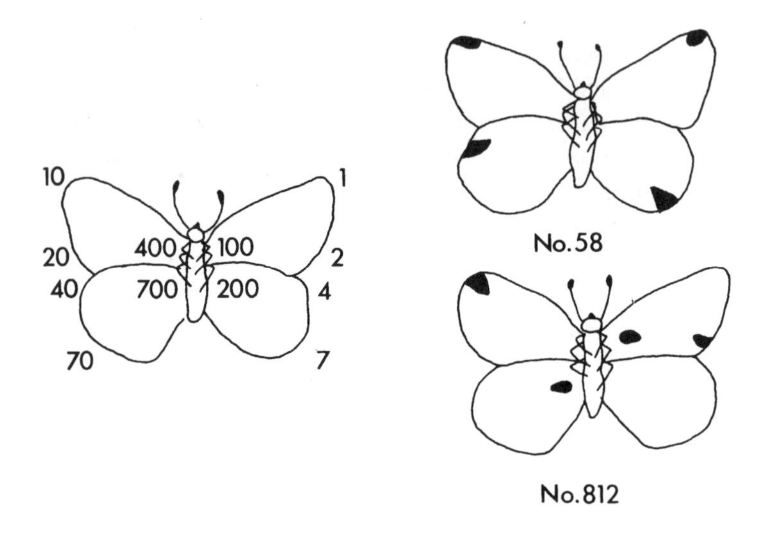

Figure 5.2 A marking scheme for the individual recognition of butterflies. The diagram on the left shows the numbers represented by marks on particular parts of the wings; two examples are given on the right (after Southwood, 1978).

trace the overall changes in the numbers and persistence of individuals within the population. The numerous indices for calculating population sizes from data of this sort are discussed in books on ecological methodology (such as Southwood, 1978; Begon, 1979), and the release of marked individuals, as well as the marking itself, may be a critical process. Wherever possible, releases should be made at the point of capture, but it is particularly important not to release a number of individuals captured over a considerable area at only one point: at the very least, they should be released throughout the area of capture to facilitate mixing with unmarked butterflies.

The underlying assumptions of methods for estimating population size may be summarized, as follows:

- Marked butterflies are essentially 'normal', without change in behaviour or life expectancy.
- Markings will persist, and not become lost or obscured.
- Total mixing occurs when marked individuals are released.
- Sampling is at discrete intervals, so that the time occupied in taking the sample is small in relation to the total time.

A simple 'Lincoln Index' also presumes that:

- The population is closed (or that levels of immigration and emigration can be assessed and accounted for), and

- There are no births or deaths in the period between samples (or, again, that allowance can be made for these).

When a single recapture occasion is used (the Lincoln Index — see Southwood, 1978 for discussion of this), the calculation is:

$$\frac{\text{Total population}}{\text{Original number marked}} = \frac{\text{Total in second sample}}{\text{Total recaptures}}$$

solved for the unknown 'Total population'.

In practice, however, multiple samples are often taken because, with an increase in the numbers of samples (occasions), the population estimates may become more accurate. Dowdeswell and his colleagues, in pioneering studies on the Meadow Brown, *Maniola jurtina* (summarized by Dowdeswell, 1982) showed how such results could be presented as a 'trellis' (Figure 5.3), representing the accumulated results of daily sampling. This method had been used in several earlier studies on various British Lepidoptera as well.

In general, some of the recent statistical innovations to MRR methods render these powerful tools for population estimation if the above assumptions can be met, and they will continue to have an important place in autecological studies of butterflies, including delimitation of territories, changes in sex ratio in the population, and so on. But quicker, if not so accurate, alternatives may be needed as a basis for ranking butterfly habitats or abundance, or where full quantitative precision is not as vital as in detailed studies of single species, or where the fate of particular individuals need not be discovered.

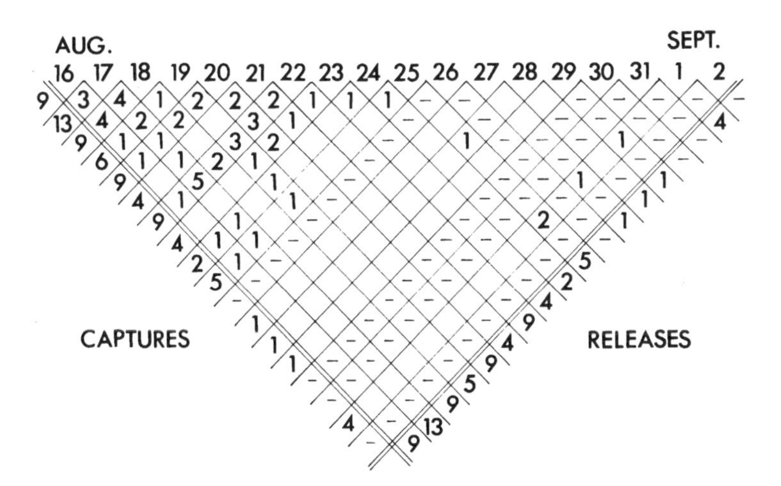

Figure 5.3 The 'trellis diagram' method for depicting results from mark-release-recapture studies of butterflies: part of Dowdeswell's data on the Meadow Brown. Successive dates are shown across the top of the trellis. Total daily samples are entered at the end of the column running diagonally to the left of the date (e.g. August 16:9, August 18:9). Figures in the body of the trellis represent numbers of recaptures of individuals marked on particular days (e.g. August 17:3 recaptures marked on August 16; August 18:4 recaptures marked on August 17 and 4 from August 16) (Dowdeswell, 1982).

DISPERSAL

To clarify vagility (dispersal) or population structure, some form of marking *is* usually required. Every butterfly study is unique and, more generally, Begon (1979) emphasized that a study cannot be properly designed without knowledge of the species or population concerned. As with other ecological studies, the results are likely to be more informative if there is a sound ecological background against which they can be appraised. As well as providing much of this initial vital information, MRR studies can help to answer specific questions about a butterfly's biology.

Scott (1975) studied and compared the flight movement patterns of 11 species of butterflies in Colorado and California. Each study site was sampled at least daily over 10–20 days, and the capture points for each individual were mapped progressively as butterflies were recaptured. There was a strong correlation between the size of the area in which the population was concentrated and the flight distances of all 11 species, and females flew further than males (in those species which yielded adequate data for the sexes to be differentiated). Most individuals moved rather short distances: of 1548 specimens recaptured, the longest recorded flight was by a male *Euchloe ausinodes* (Pieridae) which moved 2940 metres (minimum) over three days. The longest flights occurred in multivoltine polyphagous species feeding as larvae on early successional plants and, in contrast, univoltine species feeding on perennial trees and shrubs had short flights. In this example, therefore, a study of flight by MRR methods led to some tentative generalizations about habitat stability, extent of ecological specialization and distance of usual dispersal.

Vagility measurements are very important when trying to define the status of apparently disjunct populations. The Clouded Apollo (*Parnassius mnemosyne*) has declined in northern Europe and now has a highly disjunct distribution. In a study of its ecology in Finland (Vasainanen and Somerma, 1985), MRR was used to estimate and compare dispersal of the two sexes. Over five days most individuals underwent only very local movements, and occupied only about nine hectares of the total 50 hectares of available habitat. The butterflies showed a high tendency to aggregate. This, in addition to being a possible consequence of monophagy, could have survival value as defence against predators. *P. mnemosyne* are large, showy, slow butterflies which are clearly warningly coloured and distasteful to birds, and individuals probably reinforce this image by remaining close together. They are among very few butterflies recorded as 'hissing' at possible threats while at rest.

In some studies, capture of marked individuals at unexpectedly large distances from their point of release has led to a substantially increased knowledge of population structure, especially of their degree of genetic isolation. A high-altitude checkerspot butterfly, *Euphydryas anicia*, occurs in isolated colonies in alpine zones of Colorado. The larval food plant *Besseya alpina* did not occur below 3500 metres, so each colony was separated from others by inhospitable terrain on which the butterflies could not breed, and which could be considered a barrier to dispersal. However, marking of butterflies (White, 1980) showed that some inter-peak dispersal did indeed occur over a distance of around two kilometres, crossing an inhospitable area 300–450 metres lower in altitude, and against prevailing winds.

Precisely what constitutes a 'barrier' to butterflies is often not clear, but very small distances of alienated or unsuitable terrain may deter movement in some species. Although wide roads, for example, were found to slightly impede movements of some butterfly species with closed populations (Munguira and Thomas, 1992), they were not sufficient to constitute a barrier to genetic exchange. Nevertheless, roads may indeed be a deterrent to dispersal, as many butterflies have been observed turning back after beginning to cross a road.

One other relevant variable in MRR studies is the method used to initially capture the butterflies. This may be essentially random, but more structured initial sampling may be needed for some species to ensure that particular distributional subsets are not oversampled, to counter or consider diurnal activity patterns, or flight differences between the sexes, and so on. Details will vary considerably between different studies but, as Begon (1979) commented, captures can depend on the activity of the investigators or on the activities of the insects themselves — such as using baits to attract male butterflies, or introducing other unwanted biases. As we have seen, the validity of the analysis in many studies depends on the assumption that captures are random rather than consciously selected.

TRANSECT COUNTS

This apparently simple technique is now used extensively in surveying and monitoring butterfly populations and communities. It is, essentially, a relative method of very considerable value when accurate population estimates are *not* needed but trends in abundance between years or differences between sites *are* needed (Gall, 1985). It is much less expensive and intensive than MRR techniques. One merely walks steadily along a given path and counts the butterflies seen. The track taken could be along a length of roadside, a wood or forest edge, across a field or around a garden. The limitations of this method have been discussed by Frazer (1973), Pollard (1977) and Pollard and Yates (1993), amongst others, and transect counts can be used to calculate an index of abundance of the various species encountered.

In Pollard's long-term studies at Monks Wood (England), the method was standardized within limits to counter some of the influences of weather and differing activity profiles of the butterflies. Frazer (1973) had earlier emphasized that the transect survey is valuable as an easily repeatable method which, while not giving an exact estimate of population abundance, can reflect numerical changes on a day-to-day or year-to-year basis. Simple counts taken at one time of the day can be misleading because many butterflies show well-defined daily rhythms of flight activity. Yamamoto (1973) recommended making two censuses daily, at about 0900 and 1200 hours, for optimum results (Figure 5.4 gives examples of such variation).

Pollard's procedure involved starting counts after 1045 hours (British Summer Time) and completing them before 1545 hours, and assessing weather conditions. Counts were not made at temperatures below 13°C, only in sunny conditions (60% sunshine minimum) from 13–17°C and in both sunny and

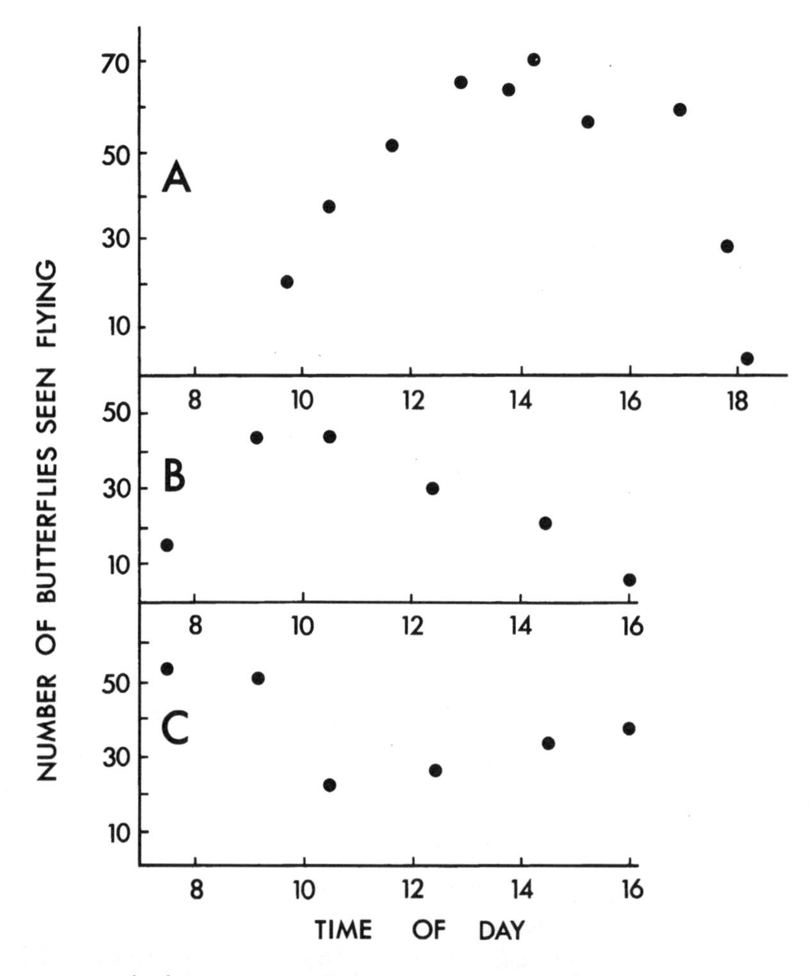

Figure 5.4 Examples of variations in counts of butterflies on a belt transect at different times of day (24-hour scale): A, *Aglais urticae* in Britain (one day, n = 465: Frazer 1973); B, *Papilio machaon* and C, *Lethe diana*, (sums for 15 days, n = 158 and 221: Yamamoto 1975).

cloudy conditions above 17°C. Counts should not generally be undertaken except in calm conditions. Transect routes were divided into sections to reflect the nature of the habitats present: separate counts from, say, different vegetation types or woodlands managed in different ways, can yield valuable information on the effects of management or change. This procedure has been adopted as the basis for transect sampling protocols elsewhere, sometimes with little critical re-evaluation to adapt it to other conditions.

The observer walks at a steady rate and notes all the butterflies seen within a definable area. Pollard (1977) noted that it is convenient to use routes with well-defined boundaries, but these can also be delimited by markers or by eye — up to about five metres in width forms a manageable area for such a 'belt transect' for which, of course, the length is also known so that butterfly abundance can be related directly to a known habitat area. Butterflies are recorded up to about five metres in front of the observer, with *minimum* numbers recorded — some butterflies will inevitably flip along in front of a walker and settle again and again, and they should not be counted more than once! If butterflies are occasionally not easily recognizable as particular species, Pollard recommended that they be recorded as the more common of the likely candidates.

The Monks Wood studies suggested that counts are reliable estimates, because several different recorders showed similar trends in the results they produced. They could also detect shorter-term changes and the duration of the flight period in any given year, in addition to year-to-year trends. For general assessment of sites, the Nature Conservancy Council suggested that each site (UK) be visited in mid-May to mid-June, mid-June to mid-July, and mid-July to mid-August, with a further period (mid-August to mid-September) sometimes advisable. A similar sequence from October–November to March–April would be applicable to Australia.

Validity of the transect method may depend, in addition to conditions noted earlier, on the following (Yamamoto, 1975):

- The area traversed must be representative of the gross habitat, and proportions of different habitats (such as open ground and woodland) need to be corrected in results before applying statistical comparisons.
- The observer must be able to recognize and identify the species without capturing them, and different observers must have a similar level of competence to ensure the comparative reliability of their information. Problems of observer bias need careful attention to ensure there is adequate standardization. Differences in individual expertise can be countered where possible, by using two observers (one expert, the other not) taking independent counts on the same transects or sites. Levels of discriminative ability may also occur so that, where possible, documenting assemblages should emphasize 'unambiguous' taxa. Some confusions can occur even for the UK fauna, such as the separation between the Small Skipper (*Thymelicus*

sylvestris) and the Essex Skipper (*Thymelicus lineola*). These are recorded simply as 'small skipper' in the British Butterfly Monitoring Scheme (Pollard and Yates, 1993). Lastly, individual observers may differ in the acuity of their observations, responding to butterflies at different distances and differing in their eyesight.

- It must be appreciated that conspicuous and/or highly active species may be over-estimated in relation to cryptic inactive ones.
- Individual comparisons between species on single samples can be made only with other species showing the same activity patterns and habitat 'preference'. The latter can be important in cases of habitat variations: a group of flowers, for example, may act as a concentration point for many butterflies of a range of species.
- Precise times, weather conditions, and duration of the survey are also useful data to be noted.

Modifications of 'basic' transect techniques may be needed to either (1) assess small habitat patches more intensively, if this can be done without damage, or (2) cover large areas with limited support. In either case — or indeed for any study — it is important to document the method in detail so that the data can be assessed realistically by other people in comparative contexts. For discrete habitats, a 'zig-zag' transect can be used as an 'area census' (Douwes, 1976) involving repeated traverses to count butterflies over the whole area. Traverses very close to others or to the habitat boundary were not recorded. At the larger scale, Mercer's (1992) survey for *Ornithoptera alexandrae* on the Managalase Plateau of Papua New Guinea involved systematic searches for butterflies, caterpillars and pupae of this large endangered species (p. 155). Roadside counts along 10.1 km of road were augmented by 21 transect lines (ranging from 183–2200 m length) at approximately 500 m intervals along the road. Large transects may be needed to adequately assess low-density butterflies, or to cover large habitat patches in determining the distribution of the butterfly. A further relatively unusual aspect of Mercer's survey was the simultaneous assessment of adults, caterpillars and pupae. The early stages were detected high in trees by using binoculars.

The intensity of transect sampling also differs between studies and for different purposes. Kremen's (1992) protocol to gain information on relative abundance of Madagascan butterflies involved transect surveys over five 10-day periods, with each of 13 transects sampled for one hour in the morning and one hour in the afternoon in each period.

There have been few attempts to compare the use of different transect methods in relation to the time and resources available for surveys. Four regimes were compared in Costa Rica for censuses of larger butterflies along 90 m transects in secondary forest (Nielsen and Monge-Nájera, 1991), as follows:

1. All individuals seen were counted within a pre-determined distance of 5 m; the distance at which each individual was first seen was also recorded.
2. All individuals seen within 5 m were recorded, but only at both sides of the observer.
3. All individuals seen at 5 m or less to the front of the observer were recorded.
4. All individuals seen within 5 m were recorded, but only to the right of the observer.

The data were derived by compartmentalizing results from the first method. Not surprisingly, the first method gave significantly higher counts than the others, as more butterflies were seen to the front of the observer than on either side. However, the last method gave results rather similar to the second and third, and may be adequate if the sampling time is limited.

As well as being used for adult butterflies, for example as a basis for detecting 'good' butterfly sites, certain species can also be appraised as eggs or larvae using a similar method. Thomas and Simcox (1982) assessed the 'nests' of the communal caterpillars of the Glanville Fritillary (*Melitaea cinxia*), which are conspicuous because of the dense silken webs they spin as hibernation retreats on their food plant, *Plantago*, on the Isle of Wight. A transect representative of the restricted habitat was selected and an observer recorded all nests within about 1.5 metres each side of his path, and the time taken to complete the survey. The number of nests was then calculated by:

$$N = \frac{wx}{y} \text{ or } \frac{wx}{z},$$

where w is the area of the site (m^2), x is the number of nests counted/site and y and z are standard rates for searching 'fast' (that is, 'flattish') and 'slow' (steep-sloped) terrain ($m^2\ min^{-1}$). This method was developed as a rapid survey technique because there was insufficient time to count all nests or, even, to search precisely measured transects or quadrats. However, the method was calibrated against the actual numbers of nests on 14 sites, determined by prolonged detailed searches, and most estimates were indeed close to the real numbers. Most species, though, do not lend themselves as readily to this treatment in their early stages; more commonly, caterpillars are solitary and cryptic, sometimes being conspicuous only at night.

Caterpillars of some other Nymphalidae bask openly, some especially in cool weather, apparently as a means of raising their body temperature. Sampling of late instar larvae formed the basis of a long-term monitoring plan for the Bay Checkerspot (p. 165) (Murphy and Weiss, 1988b). The caterpillars are conspicuous when they bask and feed on short grassland areas. Because densities vary greatly with the degree of exposure to sun, larvae were counted in samples of 50 to 100+ one-metre square quadrats in areas of relatively uniform slope exposure. Many features of the sampling regime employed reflect more general aims of such programmes, namely:

- It is non-intrusive and, as the insects are not handled, damage to them or the habitat is minimal: the technique is 'low impact'.
- It is easily repeatable.
- It gives absolute, rather than relative, population estimates.
- It documents demographic processes known to be responsible for year-to-year population fluctuations.
- It provides a baseline for future monitoring and mapping of topographic features related to habitat quality.
- It is labour-efficient, using simple approaches, and is therefore inexpensive.

Murphy and Weiss were able to demonstrate changes in populations over years (Figure 5.5), and also changes in the distribution of caterpillars as populations increased. Shifts from cooler slopes to warmer slopes affected the pattern of development and the timing of adult emergence. Comparison of the numbers found in different microclimatic regions over four years showed that initially, caterpillars were most dense on very cool slopes. All slopes showed increased abundance in year 2 (1986), but were ranked in almost the same sequence. Numbers on moderate slopes increased greatly in 1987 but almost halved on very cool slopes, and declined in all regimes in 1988 (dramatically on moderate slopes). The overall population declined from 783 000 (±81 000) in 1987 to 319 000 (±3600) in 1988, and the proportion on warm and cool slopes increased.

However, searching for caterpillars of some species is laborious and yields little information of quantitative value. Searches for larvae of the endemic Hawaiian Kamehameha butterfly (*Vanessa tameamea*) involved direct searches of foliage of the food plant, *Pipturus*; 16 360 leaves yielded only one caterpillar (Tabashnik *et al.*, 1992).

For butterflies, Pollard (1977) derived an 'index of abundance' from transect counts. The mean count per transect was calculated each week and these counts were summed over the season to give the index. If only one count was made each week, the index was simply the sum of all individuals recorded.

Pollard's 'index of abundance' is the sum of the mean weekly counts for each site and transect section over the 26 weeks from April to September inclusive. At most sites only one count each week is made, and the index of abundance is then the total butterflies seen over a season or flight period: for bivoltine species it can be calculated separately for each generation, but for species with overlapping generations (so that distinct flight periods are not evident) the counts are summed to a single annual index. As Pollard and Yates (1993) emphasized, the index assumes that variability due to the effects of weather and other factors is small compared with the effects of changes in population size, so it can be used

Figure 5.5 Bay Checkerspot (*Euphydryas editha bayensis*) habitat and population distributions in southern Santa Clara County, California in 1987 (after Murphy and Weiss, 1988). Morgan Hill is the major source population for the metapopulation; arrows denote habitat patches (all shaded areas are patches of serpentine grassland) supporting populations in 1987, those on the upper left were colonized in 1986.

as a simple basis to assess changes in abundance over many generations or years. Very few assumptions are made about the data, with the realization that it is an index of 'butterfly flight-days', a combination of butterfly abundance and longevity.

Transects can be combined with MRR techniques, so that the sampling for recaptures is based on transects and, thus, on a standardized area of habitat. Indeed, the two methods may provide a check on each other: for *Heodes vigaureae*, Douwes (1970) found a high correlation between the two kinds of estimates. Similar results were obtained for a nymphaline, using a belt transect covering the whole area frequented by the butterflies (Douwes, 1976).

EXACT COUNTS

The enormous amount of work involved usually makes it impossible to count all individuals in a butterfly population. However, for small isolated or closed populations (of colonies) this can occasionally be done — even though it may not be wise to do so because undue interference in the habitat, such as inadvertent trampling of the food plant, may increase vulnerability. It is easier for some early stages. In the case of the Glanville Fritillary larvae, noted earlier, total counts could be achieved: a good approximation could be based on knowing the number of larval nests and the *average* number of larvae in a nest.

A related species, *M. harrissi*, was studied in Canada by Dethier (1959). Larvae are confined to a single, easily detected species of food plant restricted to forest clearings, so actual counts of larvae could be made. For *Lycaena dispar* at Woodwalton Fen, England (see pp. 175–7), every individual food plant (Great Waterdock, *Rumex hydrolapathum*) was labelled and mapped, so that counts of eggs and larvae on each one could be made (Duffey, 1968). Such studies are invaluable in revealing the population dynamics of the species concerned, and as an adjunct to evaluating more restricted sampling methods.

In *M. harrissi*, for example, Dethier found that 14 200 eggs hatched, but produced only 19 adults! Many larvae died when they had eaten the food and then needed to descend to the ground in search of other plants. The few adults which emerged were, nevertheless, sufficient to produce a similar number of eggs to those of their own generation, and attempts to increase the adult population by introducing more adults from elsewhere did not succeed, as emigration then occurred. Vaughan (1989) estimated populations of the Eltham Copper (*Paralucia pyrodiscus lucida*: Lycaenidae) near Melbourne (see pp. 182–4) by nocturnal counts of larvae on the stunted *Bursaria* plants on which they feed. This has been adopted as part of the continuing monitoring protocol for this species (Braby *et al.*, 1997).

Transect counts, or other forms of regular inspection, form the basis of much of our information on seasonality and diversity of butterflies. Owen (1975) discussed three additional methods used to indicate relative abundance of butterflies at a given site. He pointed out, correctly, that transect counts may not provide an estimate of population size unless specimens are marked, because estimates will be inflated by inclusion of the same individuals day after day, especially for territorial species. These methods were:

1. Marking and releasing individuals in a particular place day after day but *ignoring recaptures* except as a means of avoiding counting the same indi-

vidual again. If this is continued for an entire season, a picture of relative abundance emerges without killing any individuals. This method is relatively labour intensive.

2. Using a Malaise trap (Figure 5.6), in which all butterflies captured are killed, is not generally recommended for conservation assessment of rare species. It can, though, provide information on diversity and seasonality with minimal effort. Butterflies and other insects are intercepted when they fly into a vertical net barrier and are 'funnelled' upwards into a tent and, eventually, into a jar of alcohol as a preservative, or into a killing bottle. Malaise traps are set up and are inspected regularly, and the collecting jar is changed at intervals (commonly, weekly). Insects are caught continuously in all weathers and by both day and night, whenever they are flying.
3. Using an attractant in baited traps (Figure 5.7), with fermenting fruit, operated in the same site throughout the season and which can be inspected once or twice a day. Trapped individuals can be marked and released as in (1). This method is especially useful for Nymphalidae in the tropics, where such 'passive' techniques may be especially welcome as an alternative to a high level of collector activity.

Especially if they are used over several years or seasons, accurate pictures of butterfly diversity and relative abundance can be constructed from these techniques, and at least some tentative information on extents of numerical change from generation to generation can be accrued. Indices of diversity can then be applied to the tabulated information. Several of these consider both the number of species present and their relative abundance, so that changes in whole butterfly faunas can be quantified. 'Diversity' is an important parameter in conservation: a 'high-diversity site' (other things being equal) will often take precedence over a 'low-diversity site' for designation as a reserve or protected area.

Commonly, though, some species are used to 'weight' this diversity; groups or ecological associations of species may constitute a 'critical fauna'; rare species,

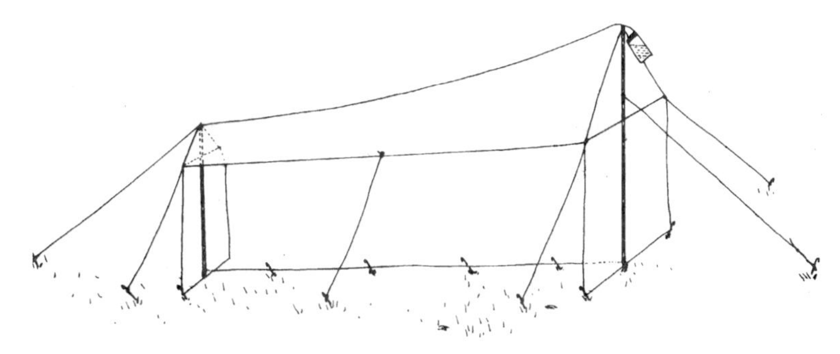

Figure 5.6 A Malaise trap. Insects fly into a vertical mesh and, as they move upwards come under a 'tent roof' and are guided upwards to a capture vessel, shown here (right) as a jar of liquid preservative. The trap is supported by poles and guy-ropes.

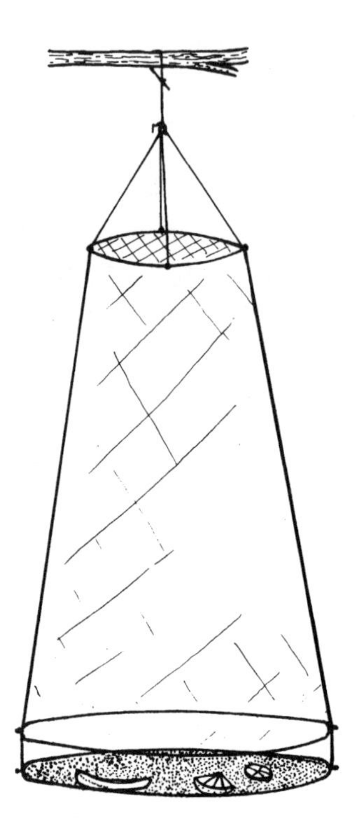

Figure 5.7 A bait trap. A platform with rotting fruit (or other baits) is suspended below a conical net, with a gap of a few centimetres for butterflies to enter. They then move upwards and are retained in the net.

those of particular evolutionary significance (such as 'living fossils' or primitive forms) and those which are independent evolutionary units (such as the only known members of their genus) may all increase the site value over a high diversity of common and widespread species. A rare species may also be in one of the notable categories mentioned. Natterer's Longwing (*Heliconius nattereri*) is the most primitive member of the South American Heliconiinae, a group which has contributed much to our understanding of mimicry systems in butterflies, and had been seen only on about three occasions this century until it was rediscovered near Espirito Santo in Brazil, by Brown (1972b). To this day it is still known from only that colony, and it is important to conserve this species because of its central evolutionary role in understanding heliconiine biology. It thus gained additional conservation priority from its scientific importance.

In less tangible, diverse faunas, such as many parts of the tropics, the assessment of assemblages may be based on some form of relative frequency. Thus, a 'DACOR' rating of butterflies of the Bulolo–Wau area of Papua New Guinea (Parsons, 1992a) recognized the following scale of sighting frequencies based on a minimum period of 10 days of observation:

dominant (D) — 21–40+ individual sightings
abundant (A) — 11–20
common (C) — 6–10
occasional (O) — 2–5
rare (R) — 1

'Rare' and 'occasional' were the largest categories in Parsons' study (127 and 134 species, respectively), with 'common' (100 species) far exceeding 'abundant' (9) and 'dominant' (2). Minor differences occurred from this general sequence in Papilionidae ('common' most frequent) and Hesperiidae ('rare' most frequent).

In assessing diversity, a parallel with ornithological pursuits is to try to obtain the 'longest list of species' from a site, even if strictly comparable and quantitative methods are not used to do this. Brown (1972a) suggested that maximizing daily species lists of butterflies could be of considerable scientific relevance — however unscientific such a goal may sound. He emphasized the value of the routine use of binoculars in helping to identify many of the larger and more conspicuous taxa, as well as high-flying species which are normally out of reach and difficult to net. These did not then need to be captured, so more effort could be put into collecting the smaller and more unusual species. In planning to maximize a daily count for an area in Brazil, Brown recommended detailed earlier reconnaissance to detect paths, streams, open areas which catch the sun at various times of the day, any high points (hill tops), flowers and other sites attractive to butterflies, as well as laying a series of bait traps on the previous evening. (A collector experienced in collecting butterflies in West Africa recommended to me many years ago that a tame goat primed with several pints of native beer to induce diarrhoea was an excellent source of butterfly bait if led along a forest track and allowed to stop at intervals to let nature take its course. Other collectors prefer to use fermenting bananas!) A collector should work from dawn to dark, although the mid-day heat may prevent activity for around three hours in lowland tropical areas. A single collector may need to move efficiently between a range of habitats, collecting in each a number of times during the day; this is where a party of collectors may be more practical and convenient, as each can concentrate on a particular area.

Brown (1972a) postulated that, using this procedure, a day's total of 350 species may be close to the practical limit for a single collector, even in extremely rich areas, because of the time involved in moving around and capturing specimens. In some very rich areas on the borders of the Amazon Basin, a group of three or four collectors might together record 500 butterfly species in a single day. Such totals far surpass any remotely possible in many temperate regions, where anything much over 30 species would be considered remarkable — even that is more than half the UK fauna!

HABITAT RANKING

Butterfly diversity, and the number of species in local communities, will continue to highlight many activities related to conservation and to be used as arguments for protection. The diverse and conflicting demands for land use, with vastly different priorities, ensure that the utopian conservation demand of reservation of all (or a large proportion of) natural areas cannot be met, and that natural butterfly habitats in many parts of the world will continue to disappear, contract, and fragment. Selection of areas for reserve status often necessitates ranking in order of desirability or suitability a sequence of grossly similar habitat patches such as woodlands, heathlands, rain forest pockets, other remnant habitat fragments, and so on. The criteria for doing this are diverse, and the literature on habitat evaluation is now enormous; see Usher (1986) for a valuable summary. The generally recognized functions of biological reserves include:

(a) conserving natural features, rare species, and communities;
(b) acting as 'storehouses' for biological evolution at both the species and community levels;
(c) being used as reference points for progressive scientific understanding and investigation of the natural world (Boden and Ovington, 1975).

Warren (1993a) established a suite of criteria for evaluating sites for British butterflies, grading sites from 'national' to 'little or no importance' on criteria of species richness and the number of species of conservation importance (Table 5.1). The population sizes of significant species was also incorporated, using the scale shown in Table 5.2, for the 29 species which were regarded as rare or declining on either a regional or national level. The aims of the study encapsulate ones which are almost universal in such work but depend on an adequate prior knowledge of the fauna:

- identify the most important sites for butterfly conservation,
- obtain accurate baseline information on the size of key butterfly populations and the condition of their breeding habitats,
- assess the coverage and effectiveness of site protection,
- recommend priority sites and species for protection, and
- recommend appropriate habitat management to ensure their survival.

As Warren (1993a) commented, conclusions from such a broad survey (which incorporated 308 sites in central southern England, within an area of about 16 000 km^2) are highly relevant to broad considerations of biodiversity conservation in fragmented landscapes. The distribution of the butterfly species revealed that many of the species breed in highly fragmented habitats which are continuing to deteriorate through neglect or unsuitable management (Warren, 1993b), with the situation no better on protected sites than on unprotected ones. For grassland species, in particular, aspect preferences gave some clear implications for conservation: most species clearly preferred south-facing

Table 5.1 Criteria for evaluating butterfly review sites (Warren, 1993a).

	Evaluation Category				
	A	*B*	*C*	*D*	*E*
Conservation importance	National	Regional	Major Local (County)	Minor Local (County)	Nil (formerly important)
Species richness					
1. No. key species	14+	10–13	5–9	1–5	0
Rarity					
2. No. RDB 1–3 species	1 large/medium	1 small	0	0	0
3. No. RDB 4 or scarce species	5+	3–4 or 2 large	1 large/medium or 2 small	1 small	0

Note: Figures indicate the number of resident species required to fulfil each evaluation category. Sites reach evaluation category by meeting at least one of criteria 1–3. See Table 5.2 for definitions of colony size. British status defined as follows: Red Data Book Species (occur in less than 15 10-km grid squares in Britain: RDB1, endangered; RDB 2, vulnerable; RDB 3, rare; RDB 4, out of danger (after Shirt, 1987). 'Scarce' species occur in less than 100 ten-kilometre grid squares in the UK (after Nature Conservancy Council, 1989). The remaining 'key' species occur in more than 100 ten-kilometre squares but are rare or declining at a regional level.

Table 5.2 Criteria for defining population size and species status (Warren, 1993a).

Colony Size or Species Status	*Estimated Adult Density at Peak Flight Period (Sightings/Hour) × Area of Site (Ha)*
Large (resident)	>500
Medium (resident)	50–500
Small (resident)	<50
Resident	Unknown colony size but breeding confirmed
Vagrant	Occasional sighting of migrants but not thought to breed
Extinct	Former colony, now extinct
Possible resident	Occasional sighting, or old records where current status is uncertain due to lack of data

aspects, which had warmer microclimates (see the discussion of *L. bellargus*, p. 174) (Figure 5.8). Some species, though (such as the Duke of Burgundy, *Hamearis lucina*) were rarely found on southerly slopes, probably reflecting the needs of their food-plants for cooler, damper conditions.

Many invertebrates are so circumscribed in distributions that any habitat patch may ultimately be unique and merit reservation on some criterion, and this is clearly impracticable. In the absence of detailed knowledge on distributions of, and threats to, restricted species, a species conservation plan *per se* cannot be undertaken comprehensively. The problem, as Thomas and Mallorie (1985) emphasized, is how to tackle the conservation of large numbers of species with specific ecological requirements, only few of which can possibly be subjects of specific investigations by the time conservation measures are needed to protect them. At present, invertebrate needs are normally considered only to a small extent (if at all) in establishing habitat use hierarchies, because their detailed biology is not sufficiently understood. As we have noted, butterflies are becoming a notable exception to this, and play a valuable role in assessment and ranking of habitats — and the two parameters noted above — diversity and the

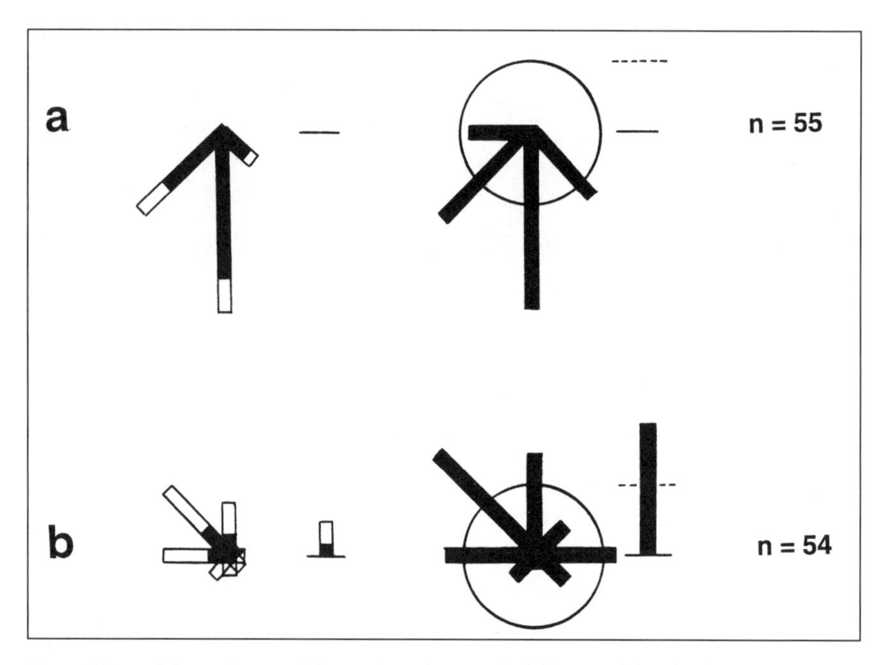

Figure 5.8 Microhabitat preferences of the (a) Adonis Blue (*Lysandra bellargus*) and (b) Duke of Burgundy (*Hamearis lucina*) (after Warren, 1993b). For each species, the polar diagram (left) gives the site aspects in central southern UK, with slopes of less than 10 degrees regarded as flat. Aspects used by large and medium-sized populations (shaded) and small populations (unshaded) are shown, together with aspect preference ratio (right: % total number of colonies per aspect / % total number of slopes per aspect). The circle denotes change of preference, with lines extending beyond the circle or dotted line showing positive preference. *L. bellargus* clearly uses and prefers southern slopes, whereas these are avoided by *H. lucina*.

presence of rare or otherwise significant species — are important facets of this selection process.

High diversity reflects ecological complexity. Thus a high number of butterfly species, or a high index of diversity, can also indicate high levels of complexity in other aspects of the community which may be more difficult to measure directly or to understand so well. Owen (1975) pointed out that a high diversity of butterflies is basically also an index of 'customer satisfaction', as visitors to a nature reserve like to see a variety of species. Rare species gain political sympathy when their restricted habitats become threatened. In such cases, the choice is clear cut: permit the change and condemn the species or population to a high likelihood of extinction, or reserve the habitat and give it a chance to survive. Here, the fundamental information needed is merely 'presence' or 'absence'. Assessment of diversity, however, may need much more extensive sampling over at least one full flight season.

Sampling diverse butterfly faunas to determine their species diversity is often by necessity restricted to short periods — perhaps opportunistic or at less-than-optimum times of the season, and (in common with most invertebrate sampling regimes) it may be difficult to determine how much sampling is 'enough'. Species accumulation curves (Figure 5.9) can indicate sampling adequacy, but they must be interpreted with considerable caution. In the example, shown (from Sparrow *et al.*, 1994, on Costa Rica butterflies), the total species curve continues to rise steadily with increased sampling effort (= number of days), indicating that only a fraction of the species present have been collected. In contrast, the curve for common species reaches a near-asymptote after only a week, so that (although the inventory is by no means complete) most readily encountered species were found with little sampling effort. Sparrow *et al.* (1994) suggested concentrating on such relatively common species, if they are sufficiently habitat-specific, to gain a reasonably rapid assessment of habitat condition. It is important to disregard cosmopolitan species likely to be common in any disturbed habitat.

Diversity was found to be higher in forest gaps created by illegal logging in northern Vietnam forests (Spitzer *et al.*, 1997), but the closed canopy habitats were of prime conservation importance because they harbour ecologically specialized species which do not invade the gaps. Many of these are also geographically restricted species.

Both detecting presence/absence and assessing diversity also benefit from information on precise habitat needs or the effects of habitat fragmentation or change — how, for instance, do different forms of woodland management, or

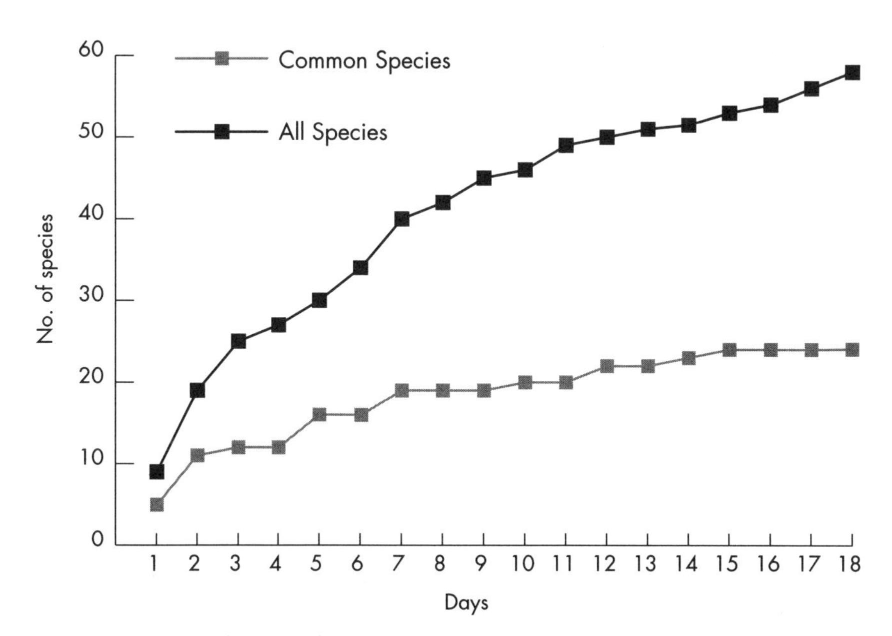

Figure 5.9 Species accumulation curves for Neotropical butterflies (Sparrow *et al.*, 1994). The cumulative number for all species (upper curve) continues to rise with sampling effort (number of sampling days), whereas the number of common species (lower curve) reaches an asymptote more rapidly (n = 114 species).

changes in the surrounds of fields, affect vulnerability and diversity? Emphasis must be on the protection of as many different ecosystems as possible, and incorporating as many 'environmental gradients' (such as altitude and geology) as practicable.

The numbers of butterfly species in any habitat are, as we have seen, assessed mainly by the presence of the conspicuous adult stage, which can be extremely seasonal. Obviously, therefore, a single sample at just one time of year can be grossly misleading, although it may be possible to differentiate well-defined phenological guilds — characteristic groups of 'spring species', 'summer species' or 'autumn species' or, in the tropics, 'dry season' or 'wet season species'. Both incidence and relative abundance of species may have strong seasonal components.

The univoltine endemic Satyrinae of south-eastern Australia, for example, show a well-defined sequence of appearance as spring and summer progress (Figure 5.10). As for butterflies elsewhere in the world, collectors and informed ecologists can predict reasonably reliably when particular taxa are likely to be flying and the times of peak abundance.

The size of the reserve may be of critical importance in relation to habitat needs. Whereas it is often claimed, in general terms, that very small reserves are adequate for insects (and this is, no doubt, commonly so) it is very difficult — even with detailed knowledge of a species' biology — to determined the *minimum* size of a suitable reserve. In general, larger areas are preferred because any minimum area must be adequately buffered around its perimeter against the effects of nearby change. Thomas (1984) summarized the information on habitat size for British butterflies: of 35 species with closed populations, 15 (Hesperiidae, Satyrinae, Lycaenidae) had minimum breeding areas of only 0.5–1 hectares, a further 11 (of the above groups, plus the Wood White, *Leptidea sinapis*) of 1–2 hectares, and only one (the forest-dwelling Purple Emperor, *Apatura iris*) needed more than 50 hectares.

Such data emphasize that, although reserving or protecting very small areas may conserve colonies (or species), destroying them could equally well destroy the colonies. Because of the imposition of (largely unknown or undetected) 'barriers' on flight activity, the fragmentation of habitat can lead to disjunctions between parts of apparently viable populations and thence to progressive genetic deterioration. This is exceedingly difficult to demonstrate, but the decline of populations to very low levels with little probability of interchange with (or immigration from) others for genetic reinforcement are causes for concern. MRR studies have demonstrated the existence of such barriers in some cases. In their work on *Maniola jurtina* in the Scilly Isles, Dowdeswell and his co-workers showed that rather small expanses of open ground could effectively isolate populations, and many workers considering the general design of nature reserves advocate the provision of 'habitat corridors' or 'stepping stones' to counter any

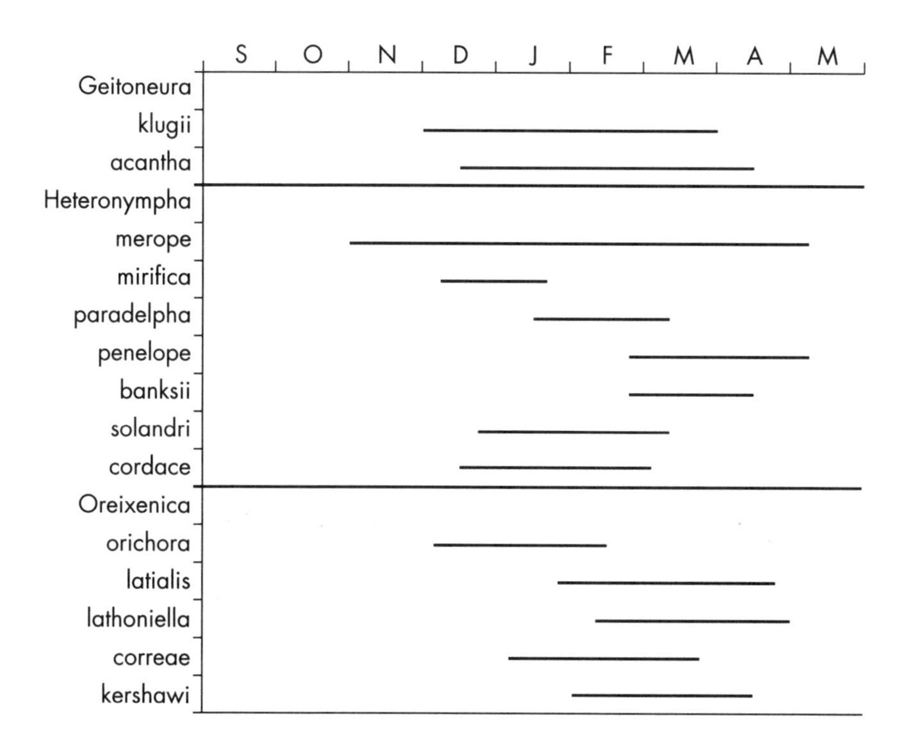

Figure 5.10 The typical flight-seasons for some Satyrinae (Nymphalidae) occurring in south-eastern Australia. Three genera are shown, with months (September to May) indicated by initial letters. Species differ clearly in their duration and the times of year they are present.

trend towards habitat fragmentation. What is usually not clear, in the case of butterflies, is whether or not such features are themselves used by or accessible to dispersing individuals. In general, land management has tended to reduce continuity between remaining sites (Thomas, 1984), and to create possible barriers between formerly continuous populations. In principle this is undesirable.

Minimum viable population sizes are also very difficult to determine — there is no room for error in experimental manipulations at the lower limits! A number of the rarest butterflies forming closed restricted populations have been assessed by direct counts. Many such colonies consist of only 50–200 individuals in most years, and Thomas (1984) noted that counts of 15–50 individuals are not uncommon, although probably often representing temporary troughs. Three species noted in Britain had declined to under 10 examples, and all were extinct by the following year. In practical terms it may be possible to prevent this by augmenting such extremely vulnerable populations by release of captive-reared insects (see pp. 129–32).

Differences in diversity between a series of habitats are also difficult to interpret due to ecological zonation. A series of ostensibly similar habitats may differ in details of topography, microclimate, altitude or vegetation which may be key resources for particular butterfly species. 'Ecological zonation' of butterflies can ensure, for example, that aspect or altitude differences can be associated with markedly different faunas. Adams (1973) analyzed collections of 501 butterfly taxa (484 species plus 17 subspecies) from the Sierra Nevada de Santa Marta, Colombia, representing samples from eight ecological zones defined on climatic, vegetational and altitudinal characters. The zones differed, not only in their total complement of butterflies and relative diversity (Table 5.3), but also in which butterflies were predominant in each. These differences would have been masked by bulking the samples to treat the Sierra Nevada as a single habitat — and the same principle applies on a finer scale to habitats with no such obvious differences. Quite what degrees of habitat differences one should attempt to categorize when defining separate butterfly communities is inevitably somewhat subjective and driven by logistics; frequently only minimum sampling intensity is possible with the resources and personnel available.

As a related example, Holloway (1976 and later papers) studied the zonation of night-flying moths on Mount Kinabalu, Sabah, by light-trap catches at different altitudes. Not surprisingly, clearly different faunas occurred at different heights. But although 'zones' can be delimited in such studies, these essentially form a continuum, so that the faunas change gradually along environmental gradients and those of any two adjacent zones tend to interdigitate. The same

Table 5.3 Zonation of butterflies in the Sierra Nevada de Santa Marta, Colombia: coefficient of relative diversity* (Adams, 1973).

	Zone							
	A	*B*	*C*	*D*	*E*	*F*	*G*	*H*
Hesperiidae	29	12	21	17	21	0	20	19
Pieridae	31	16	16	25	31	68	18	33
Papilionidae	28	20	31	7	0	0	21	5
Nymphalidae								
Danainae	152	37	16	24	16	0	31	28
Ithomiinae	0	22	23	26	0	0	22	31
Heliconiinae	0	31	26	26	26	0	25	15
Libytheinae	0	0	0	0	0	0	47	0
Satyrinae[X]	0	20	13	25	41	36	13	26
others	23	43	18	22	24	37	25	24
Lycaenidae								
Riodininae	7	10	23	12	6	0	25	10
others	9	10	22	21	10	24	13	8

[X] includes Morphinae, Brassolinae of some authors

* Coefficient of relative diversity, CRD, = nFZ/nF. nZ × 10 000, where nFZ. is the number of species of family F found in zone Z, and nF and nZ are the total complements of F (within the study area) and Z. It reflects the number of species of a family reflected in a zone if that family's and the zone's total complement of species were both 100, and allows comparisons between families and zones containing different total complements.

occurs on a broader geographical scale: the butterflies of northern Queensland and tropical Australia have much in common with those of New Guinea, while those further south in Australia do not. One priority for reserves is to typify the faunas and to represent them to the greatest possible extent.

In a particularly instructive example involving butterflies, Thomas and Mallorie (1985) assessed the effectiveness of nine different methods of selecting sites for reserves for Moroccan taxa. These differed in both the approach and the kinds of knowledge needed to assess them. They elected to choose 0–7 of 21 sites for reservation, and the success of the different methods was scored by the proportion of 'restricted species' (those with low geographical range and specific habitat requirements) and of the 41 species 'protected' in at least one designated site. The nine methods used to initially select five sites for reservation were:

1. Select sites at random (25 trials undertaken).
2. Select sites with the largest numbers of species in each.
3. Select sites with the largest numbers of restricted species (i.e. 'rarities' which are likely to be of direct conservation value if vulnerable).
4. Select sites to maximize the total number of species.
5. Select sites to maximize the total number of restricted species.
6. Select one site at random from each of five successional stages.
7. Select one site with the most species from each of five successional stages.
8. Select one site with the most restricted species from each of five successional stages.
9. Select the most dissimilar ecosystems (ignoring over-grazed and agricultural land, and omitting field margins and terraces).

Of these methods, 2 (species richness within site), 5 (species conservation) and 9 (ecosystem conservation) were compared for varying numbers of potential reserves. Methods designed to protect particular seral stages (6, 7, 8) were not particularly successful. If a knowledge of butterflies was presumed (7, 8), the proportion of species protected was lower than for other methods which assume this knowledge (3, 4, 5). If no knowledge of butterflies is assumed (6), the proportion of species protected is not much better than totally random selection (1) and is less effective than ecosystem conservation (9). For comparisons of zero to seven sites, methods 5 and 9 both accounted for all species when seven sites were included. Method 2 was less effective than 5, especially at higher numbers of sites.

Thomas and Mallorie (1985) recommended that the 'ecosystem approach' be undertaken first, with an emphasis on reserving a wide selection of environments. Following this, additional localities for other rare species could be added, after which the need for management to maintain most species should receive the most attention. This sequence is of much more general relevance and practicality.

There may be a relationship between butterfly diversity and host plant diversity in some regions, but this does not usually hold on such a simplistic level. It does hold over a range of elevations for Ithomiinae in Costa Rica (Gilbert, 1984) (Figure 5.11) and in the Atlas Mountains (Thomas and Mallorie, 1985). Spitzer *et al.* (1987) found a positive correlation also between numbers of butterflies and numbers of woody plant species in southern Vietnam savannahs. In some other communities, diversity may more represent partitioning of microclimates rather than diversity of plant species. It is therefore sometimes not clear what factors facilitate or engender butterfly diversity, except in very general or imprecise terms, but 'high diversity' is recognized as desirable. Pieridae in Morocco are distributed among different habitats and host plants in a very complex pattern (Figure 5.12).

Any geographical fauna will have associations of butterfly species which reflect the predominant habitats present — some of these can be categorized unambiguously as 'woodland', 'heathland', or 'grassland' butterflies, for example. As knowledge of their requirements becomes more complete they can be used increasingly as indicators, not merely as single species but as associations of co-existing species reflecting wellbeing of a complex suite of environmental factors.

ASSESSING DISTRIBUTIONS

Documenting changes in status and distribution is a vital facet of assessing the needs for conservation of species and their habitats. It is in this context that the masses of local lists, isolated or unusual records, food plant records and natural

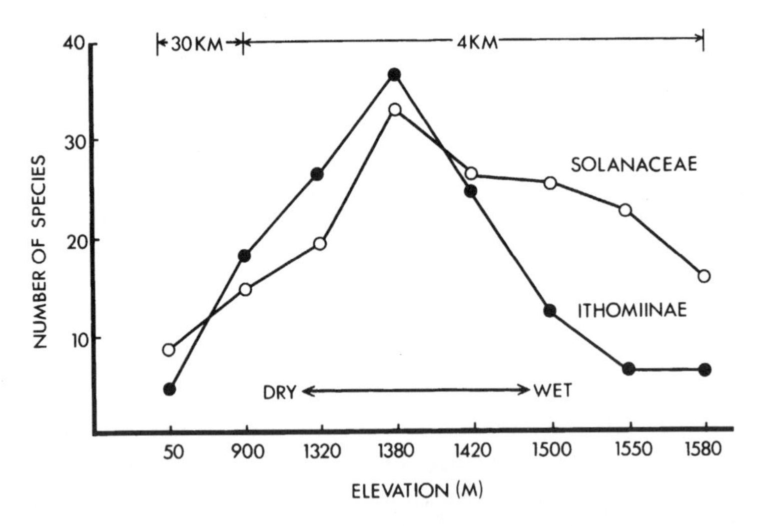

Figure 5.11 Altitudinal variations in numbers of species of Ithomiinae and their food plants (Solanaceae) in Costa Rica (Gilbert, 1984).

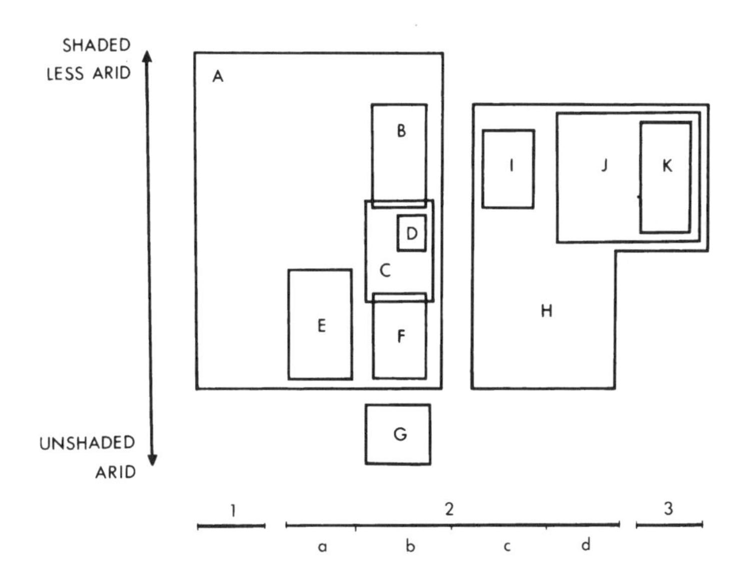

Figure 5.12 A scheme for larval host plant and habitat overlap for Moroccan Pieridae. The habitat gradient is indicated on the vertical axis, and host plant spectrum on the horizontal one: the three plant families are Resedaceae (1), Cruciferae (2), Capparidaceae (3); within Cruciferae, butterflies may feed on rosettes (a), inflorescences (b), leaves (c) or large leaves (d). The species of Pieridae are denoted by capital letters: A, *Pontia daplidice*; B, *Anthocaris cardamines*: C, *Euchloe ausonia*; D, *Zegris eupheme*; E, *Elphinstonia charlonia*; F, *Euchloe belemia*; G, *Euchloe falloui*; H, *Pieris rapae*; I, *P. napi segonzaci*; J, *P. brassicae*; K, *Colotis evagore*. Two main guilds are recognizable: leaf feeders (H–K) and rosette/inflorescence feeders (A–G) (Courtney, 1986).

history notes accumulated over the last century and more in Europe have had a substantial and vital impact as the basis for formalized recording and mapping schemes which have also spawned similar attempts elsewhere in the world. As Heath *et al.* (1984) noted, the first attempt to present information on distribution of the British Lepidoptera was as long ago as 1868 (Fast, 1868), but the great impetus to the British scheme was provided by the launch of the Botanical Society of the British Isles' distribution maps scheme in 1954.

The resulting 'Atlas' pioneered the use of a 10 × 10 kilometre square as a mapping unit for plant and animal distributions, and was adopted for Lepidoptera from 1967 by the Nature Conservancy's Biological Records Centre. The first provisional maps for butterflies were published in 1970 (Heath, 1970) and by 1982 (when this scheme ended) some 2000 contributors had submitted information from 98% of the 3600 recording units in the British Isles (Heath *et al.*, 1984). Although these records do not separate breeding colonies from casual or vagrant individuals outside their breeding range, they provide a remarkably complete set of distributional data which are envied by lepidopterists attempting to foster such schemes elsewhere, and provide an excellent basis for conservation planning. Their value is augmented by the incorporation of historical data, and three date classes are used (see Figure 5.13) and differ considerably in their mode of accumulation. Early records, pre-1940, largely represent literature records — often stretching back to before the start of the nineteenth century. For some species, this period includes known changes of distribution — expansion, contraction, or both. The second period, 1940–69, mainly covers the collections and records of early contributors to the scheme, and the 1970–82 phase predominantly comprises records submitted as they were accumulated, and sometimes deliberately sought to fill in gaps on the maps. These three major phases sometimes reveal massive changes in distribution but, as Heath *et al.* (1984) pointed out, even these may mask shorter-term changes because of the rapid decline of some species in recent years. Together, however, they indicate major trends over a considerable period.

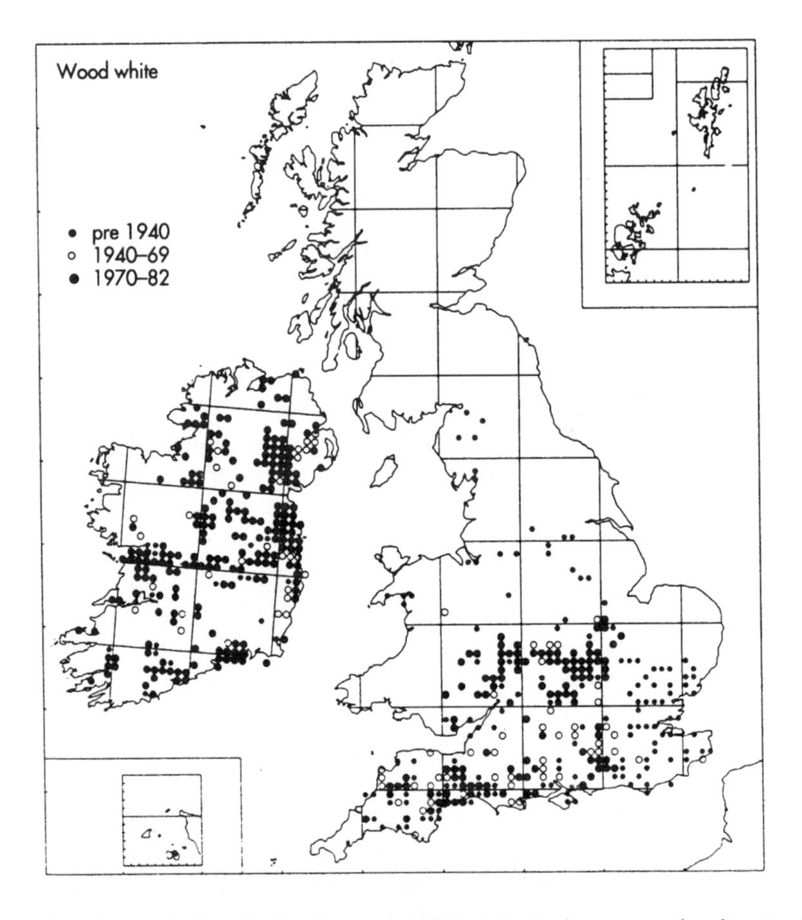

Figure 5.13 An example of a distribution map for a British butterfly, to illustrate changing range. Three date groupings are shown for the Wood White, *Leptidea sinapis*: note that the butterfly has now disappeared from much of its former northern and eastern range (Heath *et al.*, 1984).

It is, for example, very clear that the Wood White has disappeared from most of its northern and eastern England range (Warren, 1984), and that populations in central and southern England are now concentrated in three main areas (Figure 5.14). Similar contractions of range, sometimes even more pronounced, are evident in a number of other species, and most can be related directly to changes in habitat status or management. More local schemes have contributed to mapping distributions on a finer scale, such as 1 km squares, for more limited areas. In searches for patterns influencing distributions, these maps can be overlaid with those showing vegetation, weather, geology, different forms of land use, and so on.

Similar maps are gradually accruing for other parts of Europe, in particular, and are being used to estimate changes in butterfly abundance in many ways. For the Netherlands, van Swaay (1992, 1995) has emphasized the role of the long-term stability of some common species in assessing trends in the whole butterfly fauna. Species whose distributions had changed little from 1901 to 1990 were selected, and their average abundance for five-year periods was determined by calculating the average number of 5 km squares in which they were seen, as a measure of 'investigation intensity'. The number of squares for each five-year period was then calculated for other species and results corrected for investigation intensity to give the relative abundance. The common species were selected on the basis of being recorded in more than half the squares of the Netherlands; they also had to be common at the beginning of the century (present in more than 10% of investigated squares, at a time when records were somewhat sparse, with much of the country poorly or not recorded); and, lastly,

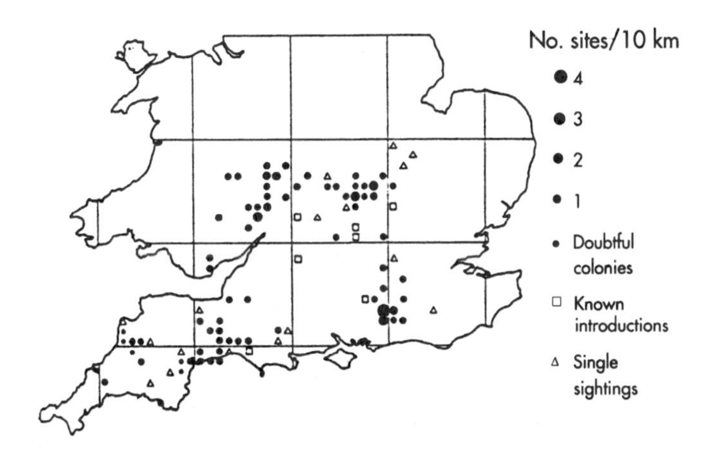

Figure 5.14 The range of the Wood White, *Leptidea sinapis*, in the UK. A further analysis, showing the approximate number of sites in each 10 kilometre square in England and Wales, 1960–79 (Warren, 1984).

they should not have fluctuated too much ($R^2 > 0.10$) or have increased or decreased notably (the direction coefficient should be between –1 and +1). These parameters led to only three species (*Coenonympha pamphilus, Maniola jurtina, Lycaena phlaeas*) being adopted as reference taxa, against which changes in other taxa could be gauged. Examples of a declining and expanding species are shown in Figure 5.15.

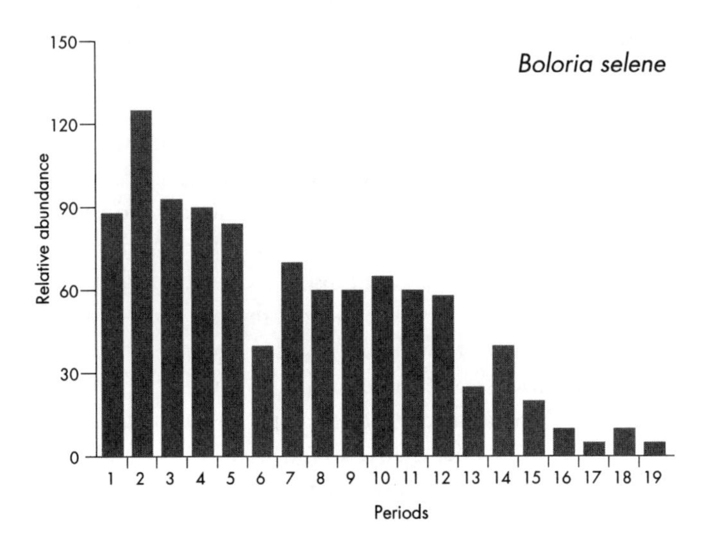

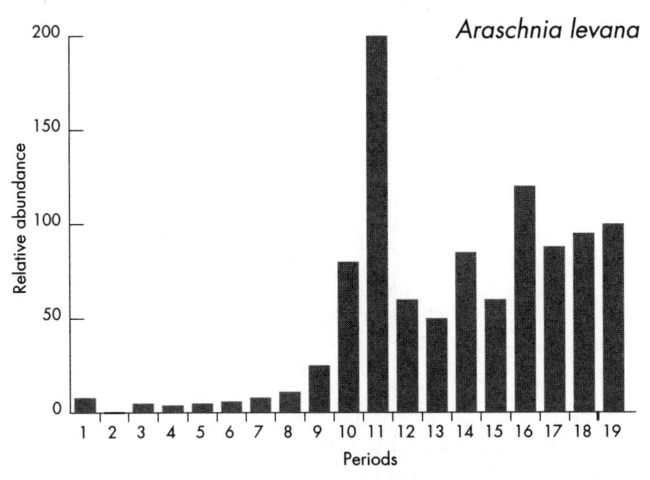

Figure 5.15 Changes in relative abundance of the Small Pearl-bordered Fritillary, *Boloria selene*, and the Map butterfly, *Araschnia levana*, in the Netherlands (after van Swaay, 1995). Each period is a five-year interval from 1901 to 1980, and relative abundance reflects the number of 5 km squares corrected for changes in abundance of selected common species to counter variations in the completeness of recording over the whole period. *B. selene* has declined considerably with loss of wet, nutrient-poor grasslands; *A. levana* is a recent colonist which has spread widely to occupy much of the country in recent decades.

Boloria selene was once widespread in the Netherlands, but depended on the wet, nutrient-poor grasslands which have been eroded steadily by agricultural improvement, so that few large populations now remain (van Swaay, 1995). In contrast, *Araschnia levana* has expanded rapidly. From a fuller appraisal of the accrued data, six categories of change could be distinguished (van Swaay, 1992):

- species which have declined continuously,
- species which have declined rapidly since the 1950s,
- species which have always been rare and declined slowly,
- species whose distribution fluctuates regularly,
- species whose distribution has changed little or not at all, and
- species which have expanded their range.

The emphasis in the UK was switched in 1976 to a 'Butterfly Monitoring Scheme', designed to monitor changes in abundance of butterflies at about 80 sites throughout the UK. Weekly counts are made along a set route (of about five kilometres) at each site from April to October, using a belt transect method and index of abundance (Pollard, 1977, p. 106). This scheme has also revealed details of flight periods for each species in different parts of the country, and some of the short term fluctuations in numbers which can result from weather and other proximal environmental influences. There are also pertinent data on status changes, what Heath *et al.* (1984) refer to as the 'prosperity' of a species. They cite, for example, that the Adonis Blue (*Lysandra bellargus* — see pp. 174–5) increased spectacularly *in numbers* from 1977 to 1982 but during the same period the total number of *colonies* in Britain declined because of destruction of breeding sites.

Four complementary schemes have collectively accumulated most of the data used for assessing butterfly status in the UK (Thomas, 1996):

1. national mapping schemes, leading to national maps based on a 10 km grid;
2. local mapping schemes, involving mainly the county scale, where intensive mapping down to 1 km squares is used to produce more precise estimates;
3. the Butterfly Monitoring Scheme, used to monitor changes in population size of most species in each generation on fixed sites, using a single transect method, and summarized by Pollard and Yates (1993);
4. rare species surveys, using a modification of the above transect method by Thomas (1983c) to involve estimating sizes of individual populations on a single visit to a site. This enabled surveys of rare species, and a single species was targeted each year to yield information on boundaries and size of every population, so that national surveys of all such species will result.

The results from each of these schemes can complement those from the others in planning conservation. However, for precise (or more precise) documentation of habitat loss, even 1 km squares are too coarse, and Hardy (1994) investigated the use of 100 × 100 m squares to determine the existence of particular biotopes and the butterflies which depend on them.

In the UK a comprehensive atlas of butterfly distribution is projected for publication under the banner of 'Butterflies for the New Millenium' (Butterfly Conservation, 1996). The major aim will be to record and map butterfly colonies, using a grid of squares (10 km, 2 km or 1 km sides) for various purposes. Where a site is noted, it is recommended that its location be recorded on at least a 1 km square grid, and preferably on a 100 m square grid, based on Ordnance Survey maps. This level of detail can only be achieved with the aid of numerous informed observers and enthusiasts, but will form a useful template against which to assess future changes.

In the UK around 200 separate butterfly recording schemes have been instituted, with varying aims, scope and rationale (Asher, 1992). Some are general, some local; some target particular habitats (such as gardens or churchyards); some extend over only one season, others are long-term; but virtually all of them depend on the enthusiastic participation of volunteers and, as Harding *et al.* (1995) noted, 'without their contribution our present knowledge would be inadequate and any plan for future recording will be almost entirely dependent on their continued commitment'.

The objectives of recording schemes fall into five main groups (Asher, 1992):

- provide knowledge of the distribution and status of a butterfly fauna,
- identify sites important to butterflies and their conservation,
- identify changes in the wellbeing of butterflies at particular sites and more broadly,
- assess the factors causing change, and
- provide a basis from which to advance conservation priorities and advise on planning and legislative matters.

Compare the completeness of the British records with those from Victoria, Australia, an area similar in size to Britain, arguably the most intensively collected mainland state of Australia but still one where butterfly collecting has always had only a small number of devotees. Figure 5.16 (p. 127) shows those parts of Victoria, divided into 10 × 10 minute areas (about 15 × 18 kilometres) from which *any* butterflies had been recorded up to 1985 in the Entomological Society of Victoria's ENTRECS scheme. Much of the state is still blank, although many further records have accumulated during the last few years, but the lack of tangible data equivalent to that for the UK has hampered attempts to conserve some rare butterflies, as politicians tend to take the lacunae as representing lack of concern rather than logistic limitations. For much of Australia, the gross distributions of butterflies are known and (as for equivalent handbooks elsewhere), the maps in Common and Waterhouse (1981) are gradually being refined as new information accumulates and are a good basis for at least detecting

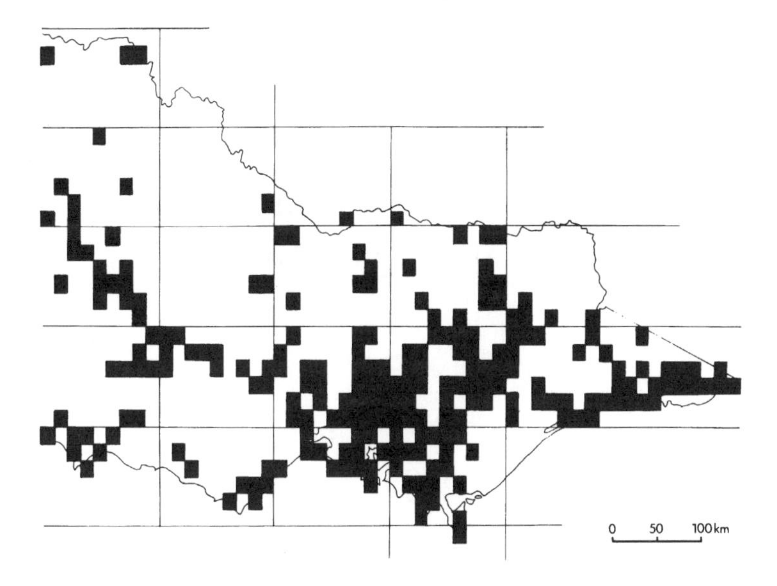

Figure 5.16 Knowledge of butterfly distributions in Victoria, Australia: the black areas, each of 10 × 10 minutes, represent the areas for which *any* butterflies had been formally recorded by the Entomological Society of Victoria's ENTRECS scheme by April 1985. The situation has been improved substantially over the last decade, but the map shows well the lack of good historical information on butterfly distributions in the region.

species outside their known range. The Victorian scheme does, however, exemplify the progress which can be made without any 'official backing', and stresses again the vital role of non-professional workers.

In order to partially formalize the accumulation of information on butterfly distributions in North America, the Xerces Society initiated an annual 'Fourth of July Butterfly Count', which has continued to the present under the auspices of the North American Butterfly Association. For about a month centred around mid-June to mid-July each year, butterfly enthusiasts visit sites and record the identities and abundance of butterflies they see. Some sites have now been visited regularly over several years, and some changes of status have been revealed. Notes are made on vegetation, weather conditions, duration of the observations and changes of status of the land since any previous visit. The results, published each year, are assembling an enormous body of information. Two hundred and sixty-six people participated in counts at 46 sites in 1982, when 346 species were recorded; by 1986, 64 sites were assessed, and by the early 1990s more than 200 counts were being conducted in more than 30 states, five Canadian provinces and Mexico (Opler, 1995).

CHAPTER SIX

TOWARDS MANAGEMENT OF BUTTERFLIES OR HABITATS

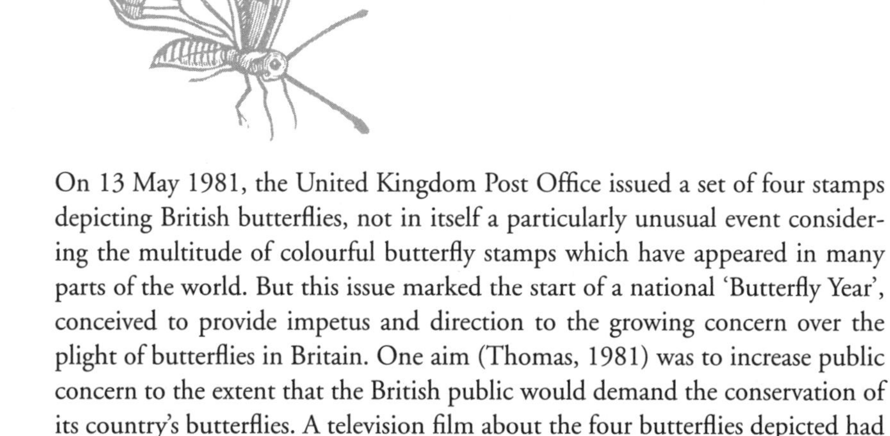

On 13 May 1981, the United Kingdom Post Office issued a set of four stamps depicting British butterflies, not in itself a particularly unusual event considering the multitude of colourful butterfly stamps which have appeared in many parts of the world. But this issue marked the start of a national 'Butterfly Year', conceived to provide impetus and direction to the growing concern over the plight of butterflies in Britain. One aim (Thomas, 1981) was to increase public concern to the extent that the British public would demand the conservation of its country's butterflies. A television film about the four butterflies depicted had been first shown some months earlier and was watched by more than three million people, and many projects throughout the year constituted a major attempt to publicize the needs of butterflies, including fund-raising for specific conservation projects. One of the bodies which provided major impetus for the Butterfly Year was the British Butterfly Conservation Society, with a membership of around 1800 at that time, but which (as Butterfly Conservation) has grown greatly since then. The Society had the following aims (Tatham, 1981):

- To save from extinction or protect all species of British butterflies, by conserving them in the wild by such means as are available, or by breeding numbers in captivity and, where possible, reintroducing them in natural habitats;
- To sponsor further scientific study and research in conservation of these butterflies both in the wild and in captivity;
- To foster interest generally by educating the public, and in particular educational establishments, in problems concerning conservation of the butterflies.

Together, these objectives summarize much of the major thrust of modern butterfly conservation, based on sound ecological study augmented by captive rearing (when needed) and education. They also introduce us to several topics which need more discussion.

CAPTIVE REARING

For species-orientated conservation, 'captive breeding' and 'reintroductions' are especially important issues. The basic rationale is that if a species is rare or threatened in nature it might be possible to rear large numbers in captivity, protected from some of the usual factors causing mortality and thereby increasing survival, and use this stock to augment field populations — either the population from which it originated or, more controversially, to reintroduce the species to areas from where it had disappeared or to completely new sites. A related topic, butterfly farming, may also involve captive breeding, but usually for rather different motives. Satisfying collector demand for rare species (pp. 62–3) by supplying these from captive stock rather than by captures from shrinking natural populations can only be of positive conservation value. One of the first major attempts to do this occurred in Papua New Guinea.

THE PAPUA NEW GUINEA SCHEME

In 1966, the Papua New Guinea Government listed seven species of rare birdwing butterflies as protected, with heavy penalties imposed for killing, trading or exporting them. The species were *Ornithoptera alexandrae*, *allotei* (now known to be a hybrid form), *chimaera*, *goliath*, *meridionalis*, *paradisia* and *victoriae*. A booklet depicting these 'National Butterflies' in full colour and at natural size (Mitchell, n.d.) was produced to facilitate their recognition. In parallel with this legislative protection, the Government formed a central insect marketing board, the Insect Farming and Trading Agency (IFTA) at Bulolo, Morobe Province (now allied with the Papua New Guinea University of Technology, Lae). This was designed to foster a cottage industry based on local, mainly unendangered, insects (Pyle and Hughes, 1978). Only Papua New Guinea citizens may sell native insects, and the IFTA is the only legal outlet in the country for these. Emphasis was placed on rearing insects for sale, as well as capturing them for trade, and the IFTA has the long-term aim of being able to supply collectors with perfect specimens of any species of butterfly in Papua New Guinea (Parsons, 1978).

Many of these are projected to come from rearing or farming operations, and the main butterfly species targeted in the early stages for commercial farming were two common birdwings, *Ornithoptera priamus* and *Troides oblongomaculatus*. Attractant plants (p. 188) and larval host plants are used to concentrate offspring of wild females, and a proportion of the reared specimens are sold, others being released. A farming manual (Parsons, 1978) gives detailed instructions to local people on how to establish butterfly farms of about half an acre. A surrounding hedge of *Hibiscus* and other nectar-bearing plants provides food for adult butterflies and keeps out pigs, and the enclosed area contains a common vine (*Aristolochia tagala*) on shade trees or others and other food plants. Pupae are collected from the food plants and protected from ants and other predators while the birdwings emerge. Instructions are given for killing and papering the butterflies for transmission to the IFTA. Collecting of other species is also encouraged, but it is stressed that only perfect specimens should be taken for sale, and that butterflies should not be over-collected in any area.

This programme was one of the first, especially in the tropics, to realize the importance of butterflies which are desired by collectors as a rationally harvestable and indefinitely exploitable resource. It has led to the preservation of selected habitats both locally and on a broader scale, rather than unthinkingly clearing all native vegetation and, although the protected birdwings have not yet been farmed in this way, it may eventually be found feasible to do so.

The Papua New Guinea scheme expanded rapidly. It commenced with the participation of about 30 people in two provinces in 1974, but four years later more than 500 villagers in 10 provinces were farming or collecting butterflies in conjunction with IFTA, and this number continued to increase (National Research Council, 1983). The cultural change engendered by this scheme is also important. By 1981 it was estimated that a butterfly farmer could earn about US$1200 a year. This is a substantial income in rural areas of Papua New Guinea, where the mean per capita income was then in the order of only US$50 a year. Many people in remote areas were able to participate in a cash-based economy for the first time in their cultural history. By the early 1990s, Orsak (1993) believed that incomes could approach the minimum rural wage, which few people earned because of high unemployment. At the other extreme, up to US$14 000 annually was noted as income from farming valuable island butterflies.

This scheme has acted as a model for others in various parts of the world, where species coveted by collectors are progressively being targeted for ranching operations. It is recognized that some form of 'trade protection' may be necessary; in Papua New Guinea it is illegal to export living insects, which is also a very effective way to ensure that breeding stocks cannot be established elsewhere!

This is a different approach from that taken by suppliers of livestock for butterfly houses (pp. 65–7), and some other butterfly farms projected for development in the near future give greater priority to this fast-growing trade opportunity. Whatever the emphasis of a particular operation, butterfly farming has an important role to play in conservation, as well as providing local employment as a 'cottage industry' with considerable growth potential. However, the major impetus for development of commercial butterfly farms in the short and medium term is likely to be for the provision of showy, moderately rare taxa. Programmes to commercially exploit butterflies 'must either bypass the rarest, most restricted and slowest reproducing species; farm them cautiously with a high percentage of adults released, or take them with the greatest of care' (National Research Council, 1983).

The dilemma of conservation versus development is paramount in many 'less developed' countries, and butterflies are increasingly important as a utilizable, sustainable resource — though as Parsons (1992b) pointed out, they are of no use to people in their prime needs of food, fuel and shelter. The challenge remains to determine if butterfly farming can become truly effective on any broad scale in preventing tropical forest destruction, by increasing people's income from secondary forest systems and other systems which can be enriched to form butterfly gardens or attract commercially viable species.

Orsak (1993) has emphasized that fostering trade to collectors and others in birdwing butterflies and other desirable species is a vital aspect of practical conservation in Papua New Guinea, by fostering conservation through development. The idea of buying dead butterflies, though it may be increasingly repellent to butterfly enthusiasts in countries where the activity of collecting is becoming socially abhorrent, is a critical aspect of increasing practical conservation over parts of the Indo–Australian region (Parsons, 1992b, 1996a; Orsak, 1993; New, 1994). As Orsak put it in relation to Papua New Guinea, 'butterfly business' promotes an especially strong link between conservation and development. In essence 'butterfly well-being leads to people well-being'. That is a powerful connection and incentive for conservation. PNG is well placed for expansion (Mercer, 1989), and the initial aims of IFTA included ensuring that insects are treated as a renewable resource and promoting the conservation of butterflies and their habitats, together embracing the overall rationale for the system (Parsons, 1992b; 1996a). But only through such centralized control can a benefit to local people and the environment be assured. Illegally captured and traded butterflies, probably a very high proportion of the rare species listed in dealers' catalogues, merely result in profits to the dealers and agents, rather than to the supply level.

Parsons (1996a) noted the need to avoid local domination of trade activities by 'collector barons', and listed the important rationale for a central control agency in controlling quality, in maintaining the confidence of suppliers in assuring that they are paid fairly and promptly, as an educational centre for farming and trading methods, and as suppliers of equipment, in addition to coordinating associated research and monitoring.

As an example of a situation without such control, Hoskins and Hardy (1993) noted the extent of the captive and trade market in Rajah Brooke's Birdwing (*Troides brookiana*) in Malaysia, where children use dead or crippled butterflies as decoys to attract others. Hoskins and Hardy were sceptical about the value of local butterfly farms as having any role in scientific or conservation advancement.

Butterfly farming operations are similar to any other business enterprise, with logistics a prime concern in deciding whether or not to establish them. The time that a product takes to reach the market may be especially critical for livestock, so that remote areas far from good international air services may not be suitable. An inventory of local taxa may be needed in order to select a range of species to provide and concentrate on. Possible liaison with agro-forestry operations may provide benefit, as may collaboration with tertiary research or educational institutions; and local tourism and employment benefits may also be relevant.

Many of the attempts in the Indo–Australian region were discussed by Parsons (1992b, 1996a). Operations in other parts of the tropics (such as Central America) also exist, though some of these are private operations, so that the extent of the conservation benefit is at the behest of the individual proprietor whose main interest may be commercial rearing to supply butterfly houses elsewhere in the world (Ross, 1996).

This is clearly very different from rearing solely for reintroduction. In this case, rearing is concentrated on particular species of conservation interest, which does not aim primarily to satisfy commercial demand (though this may well be considered) but to augment the size or number of field populations. It is therefore costly.

REINTRODUCTION

The principle of 'reintroduction' is a controversial topic, now recognized as a valuable facet of many conservation programmes. It has a substantial history, with attempts ranging from introducing exotic species (such as the Map Butterfly, *Araschnia levana*, from Europe into England around 1912, where it survived and increased for several years before being subsequently exterminated by a collector who objected to the introduction of an alien species) to a more-or-less regular adjunct to conservation management. It has thus been used predominantly to augment small populations or to reintroduce species into areas from where they have recently disappeared and for which the reasons for the disappearance are at least partially understood.

Quite a lot of introductions, either inadvertent or deliberate, have not been well-documented in the past, and so it is not always clear whether they have occurred or whether some species have expanded their range naturally. In the UK, the White Admiral (*Ladoga camilla*) underwent a remarkable range expansion in the 1930s and 1940s (Pollard, 1979) and Ford (1945) noted that this natural expansion included areas to where it had been introduced without success early in the century (Figure 6.1). There have been several attempts to

Figure 6.1 The range expansion of the White Admiral, *Ladoga camilla*, in the UK. The shaded areas include the vice-counties which yielded records from 1901–10 (left) and 1941–50 (right) (Pollard, 1979).

reintroduce the Blackveined White (*Aporia crataegi*) into the UK, where it became extinct about 1925. The general lack of success has been attributed to bird predators and the rather extreme climatic conditions on this northern fringe of its former European distribution (Pratt, 1983). An introduction of *A. crataegi* from Spanish stock to Scotland in 1974 needed protection from these factors in order for it to persist.

An often-implicated source of introduction is the release of surplus reared stock by collectors, and the clear responsibility here is to liberate only in the region where the parents or caterpillars were earlier found, rather than 'just anywhere'. A Xerces Society policy (Pyle, 1976) comments that the introduction of live Lepidoptera as a conservation measure is 'extreme, potentially dangerous, and not to be undertaken without consideration of possible effects'. The origins of *exotic* species are not always clear, and a number of early cases were discussed by Ford (1945), but any attempts to spread *native* species do need to be very closely documented so that they are not confused with natural changes in distribution; this need was recognized and recommended long ago (Sheldon, 1925).

Very occasionally introductions may be considered the *only* solution to a conservation problem (see pp. 175–7), and Pyle (1976) recommended that these should only be undertaken then if:

- the native population is known to be extinct, or
- the introduced stock is genetically similar (from the same population/s) to any native individuals which may still be present, and
- introduction takes place within the natural range of the species concerned.

The term 'reintroduction' is commonly used for any introduction of a butterfly into a former or a 'new' area. However, there are several rather different contexts involved here.

The JCCBI (1986) has also issued a Code for re-establishment, in which several categories of live releases are recognized. These differ in their emphasis and are noted here to demonstrate the various practical contexts in which releases may be contemplated. They are:

- 'Re-establishment' is restricted to the context of a deliberate release and encouragement of a species in an area where it formerly occurred but is now extinct. It is recommended that no species should be regarded as locally extinct unless it has not been seen there for at least five years.
- 'Introduction' means attempting to establish a species in an area where it is not known to occur or to have occurred previously.
- 'Reintroduction' entails an attempt to establish a species in an area to which it has been introduced but where that introduction has not succeeded.
- 'Reinforcement' means attempting to increase population size by releasing additional individuals into the population.
- 'Translocation' means the transfer of individuals from an endangered site to a protected or neutral one.
- 'Establishment' is a neutral term used to denote any artificial or intentional attempt to increase numbers by transfer of individuals.

The JCCBI Code emphasizes that re-establishment can be either species-orientated (aimed at endangered or vulnerable species threatened with habitat destruction) or site-orientated (enhancing the value of a site by providing notable species which were formerly there), and also stresses the importance of assessing the suitability of the site to support the introduced insects. The source of the introduced stock may also be important and, in general, that from a nearby locality is preferable to more distant stock.

In the UK, Butterfly Conservation has attempted to bring terminology for butterfly reintroductions (which the organization prefers to term 'restoration') into line with the standard terms used by the World Conservation Union, as follows:

Introduction — The intentional or accidental dispersal by human agency of a living organism outside its historically known native range.

Reintroduction — The intentional movement of an organism into a part of its native range from which it has disappeared or become extirpated in historic times.

Translocation — The movement of living organisms from one area with free release in another (Conservation Committee of Butterfly Conservation, 1995).

They stress the need to place such operations in the broader contexts of a Species Action Plan (p. 87) and the wider conservation of the species.

Reintroduction is thus a practical option when 'essential priority measures concerning habitat and species protection, management and monitoring are being implemented as far as possible' and 'a species, subspecies or race has become extinct, or is threatened with extinction through loss of sites or long-term decline at (a) a national level; or (b) a regional level; or (c) at genetically or ecologically important sites'.

A series of caveats and conditions for consideration when formulating such a strategy (see Appendix 2) are designed to foster high levels of responsibility and documentation, and to deter 'casual' releases and movements of insects.

Because of the need for occasional cases of augmentations of small populations on a more or less regular basis, Morton (1983) suggested the formation of a 'Captive Breeding Institute' for butterflies in the UK, with the prime aim of maintaining stocks of such rare taxa for use in establishment programmes. He emphasized the greater efficiency of doing this from a central facility rather than through a series of separate isolated operations. Recent trends towards the use of artificial diets for feeding caterpillars (pp. 195–6) could be useful in large-scale rearing, but some of the longer term risks, such as genetic deterioration through persistent or prolonged inbreeding, are extremely difficult to evaluate.

A released population must be carefully monitored in order to determine its fate, and the reasons for this. Some butterfly introductions have been spectacularly successful, in contrast to the failures noted earlier. The introduction of the Heath Fritillary (*Mellicta athalia*) in one UK site led to the survival of the colony for about 20 years, until woodland management rendered it unsuitable as a habitat. One of the more famous British reintroductions is of the Large Copper (*Lycaena dispar*) to replace the endemic subspecies which became extinct in the last century (pp. 175–7).

A study of the ecology of the Swallowtail, *Papilio machaon*, was initiated in the UK in 1971 in the hope of assessing the likelihood of introducing it to Wicken Fen, where it became extinct in the early 1950s (Dempster and Hall, 1980). Fifty eggs were collected from another colony and used as the foundation for rearing captive stock, and 228 adults (124 females, 104 males) were released in 1975. Virtually every large food plant received eggs, with an estimated 20 000 deposited throughout the fen. Over 2000 caterpillars survived to pupation. More than 100 adults were seen in 1976, but only three in 1979, and it seems that no eggs were then laid and the Swallowtail again became extinct at Wicken after only four years. Dempster and Hall (1980) found no direct evidence for inbreeding decline as a cause, although the food plant became progressively more 'patchy' over the period, and it was suspected that the Fen needed to be maintained in a wetter condition to foster food plant (Milk-parsley, *Peucedanum*) wellbeing. Rotational cutting of sedges every three to four years has led to food plant decline, and patchy populations are susceptible to accident by unintentional cutting. As in many other cases of butterfly conservation, the management of the habitat to safeguard food plants is a central issue.

As a result of recent (1988–90) management to reconstruct the banks around Wicken Fen, it is now much wetter and the decline of *Peucedanum* has been reversed. Dempster (1995) reported another attempt to re-establish the Swallowtail there in 1994, and this range extension is included in the Species Action Plan (Barnett and Warren, 1995b). Butterflies were seen there in 1995, but there were few offspring, possibly because of the loss of young food plants during a summer drought (Dempster and Hall, 1996).

Before discussing management further, there is one other 'captive breeding' context which should be mentioned at this point, and this concerns the live exhibits of the butterfly houses noted earlier. Butterfly houses can be very much more than simply large cages of living insects. They have a substantial scientific role to play in documenting butterfly biology and in acting as interpretative centres for the public. Some now produce very informative printed guides and sophisticated educational materials, as well as being providers of research material. Some undoubtedly have the potential to produce surplus stock of some species which could be released, and the provision of livestock for conservation (as initiated in New Guinea) is an important, though not yet widespread, adjunct function of some butterfly farms or houses. Anyone interested in butterfly rearing (Chapter 8) can obtain valuable tips from, for example, looking at the methods used to feed adult butterflies in these live displays, as they may lend themselves easily to adaptations useful in the home garden in order to encourage butterflies there (p. 190).

SITE EVALUATION AND MANAGEMENT

Habitat preservation, albeit a vital part of protecting butterflies against many forms of environmental change, is often only the first step in practical conservation. It is commonly not sufficient merely to provide a habitat and to assume that all will then be well in perpetuity. More often, active management of resources may be needed to facilitate or ensure this. The examples outlined in Chapter 7 emphasize this point in a range of different contexts, but also show that the actual acquisition of habitat may be both expensive and very difficult. The form of management undertaken will then influence the carrying capacity of the habitat, so that the value of small habitat areas may be considerably enhanced.

We noted earlier that butterfly diversity and habitat area are among the factors used commonly to evaluate possible reserves. At present, most reserves must be founded without a detailed or sufficient knowledge even of particular rare butterflies which they are meant to protect, and it is often necessary to determine the smallest area which will suffice. Intuitively, many people feel that larger reserves are more desirable than smaller ones, because of a direct relationship between numbers of species and area, but there has been much discussion on how a given area should be allocated: the problem is sometimes summarized as SLOSS (Single Large Or Several Small reserves).

In an important study on reserves for butterflies, Shreeve and Mason (1980) assessed the butterflies of a series of woodlands in the UK and found that the number of species present was indeed positively correlated with habitat area. Directly extrapolating from their data, they estimated that 458 hectares were needed to support the 26 species of woodland butterflies. However, one wood of 85 hectares actually contained 22 species, and direct addition of areas of representation gave a practical *minimum* of 180 hectares for all 26 species to be present. The 85-hectare wood contrasts with one nearly twice as large but with only 18 species, none of which was considered rare, whereas the smaller wood harboured five rare species. In practice, theoretical generalities of this sort should be questioned and, although the detailed knowledge necessary to select key sites for insect groups usually does not exist, one aim of conservation assessment is to acquire and rationalize this as well as possible. Specialist butterflies, often with closed populations, may be particularly informative as site indicators. Pyle (1982) found that a number of Washington butterflies had high value in reflecting site uniqueness — with the thesis that knowledge of distribution should indicate the rationale for nature reserve planning to retain regional diversity. Pyle recommended that butterfly conservation could here be approached initially from a biogeographical viewpoint, then through specific ecological investigations which would lead to development of management plans.

The UK 'Butterfly Site Register' scheme (Warren, 1985a) suggests that priority sites for a species will fall into one of two categories, each with several features:

1. Good existing sites, whose features include (i) good representation of the selected 'key' species, which will vary depending on habitat and, in many cases, geographical region; (ii) viable populations of the key species, so that sites will need to be sufficiently large for this, and (iii) some wider conservation value.
2. Sites with potential for becoming good sites, whose features include (i) substantial historical representation of the 'key' species; (ii) potential for providing suitable habitat and viable populations if the management was improved; (iii) sympathetic landowners; (iv) reasonably large size, and (v) potential for establishment of some of the key species if management was improved and natural colonization was unlikely.

Kudrna (1986), in his monumental work on butterfly conservation in Europe, derived a 'chorological index' for each species and, as Viejo *et al.* (1989) noted, these can be averaged for the species present at a site to reflect the value of the locality. Kudrna's index consisted of summing the following values for each species. They reflect three different, but complementary, aspects of distribution:

Range size — a measure of the area of the species' distribution. Five values are designated:

- widespread over the whole, or nearly all, of the area (in Kudrna's context, of Europe);
- species widespread over large parts of the area;
- species localized to one or more smaller parts of the area;
- species restricted to one or more areas smaller than the above;
- species confined to one small area, such as a single island or site.

Range composition — reflecting continuity of distribution as the ability of individuals in one colony to reach those in another. Again, five values are designated:

- continuous throughout the continental range;
- widespread over most of broad range but with small proportion of relatively isolated colonies in some areas;
- colonies which tend to be isolated but with a good proportion of continuous distribution in central parts of the range;
- discontinuous distribution in scattered colonies over nearly all the range;
- widely separated disjunct single colonies or groups of populations.

Range affinity — the relationship between the species' continental distribution and its world range was considered to supplement the overall significance of populations. Four values were defined:

- species poorly represented in relation to elsewhere;
- 'neutral' species widespread both within and without the area;
- species with their centre of distribution within the area but known also from other continents;
- endemic species not found elsewhere.

This scheme, with minor modifications, is potentially useful elsewhere. The category sequences in these lists are translated directly into numbers for relative quantitative appraisal. In Australia, *Pieris rapae* would have a chorological index (range size + range composition + range affinity) of (2 + 1 + 2) = 5, compared with (1 + 1 + 2) = 4 in Europe; *Heteronympha merope* (p. 36) would, subjectively, sum as (2 + 3 + 4) = 9, and many restricted taxa (those of high conservation value) would score more highly: the Australian Hairstreak (p. 39), and the Eltham Copper, (p. 182) for example, both as (4 + 5 + 4) = 13.

In their study of Spanish woodland butterflies, Viejo *et al.* (1989) utilized this index along with several other parameters — the number of species, the diversity (in this case, the Shannon–Weiner diversity index) and the 'average biotope amplitude': an index of 'status' defined as the total number of sites where the species was present divided by the number of sites surveyed, thus reflecting 'rarity'. Particular kinds of woodland clearly supported richer and more significant butterfly faunas than others, and these merit preference for reservation or management.

It may be feasible to categorize butterflies in local faunas by the extent of their dependence on specialized habitats. In formulating the general needs for conservation for rare Lycaenidae in Spain, for example, Munguira and Martin (1993) recognized three categories of habitat dependence:

1. Species living in habitats with climax vegetation. These, such as *Lysandra golgus* and *Agriades zullichi* (see also Munguira *et al.*, 1993), are high-altitude species which live in areas where past human impact has been low. Nature reserves, such as National Parks, may be the best option in such areas, to counter the increasing pressures to develop areas such as the Sierra Nevada for tourism and winter sports.
2. Species living in early successional vegetation on natural rocky outcrops and scree. Management needed to conserve such habitats is minimal. Such areas may be maintained naturally by landslides or avalanches.
3. Species dependent upon habitats kept open by traditional land uses such as grazing, mowing and burning. These exemplify some of the regimes that are also of great concern elsewhere, as losses of populations are often due to more intensive agriculture or to abandoning traditional current practices. The main conservation need hinges on the recognition that such practices are an integral and necessary part of sustaining habitats suitable for many characteristic species (see also Erhardt, 1985; Erhardt and Thomas, 1991). Many such butterflies respond rapidly to either form of departure from traditional use of grasslands, invariably reflecting their narrow specialization, such as depending on particular sward heights for oviposition, or for the wellbeing of a mutualistic ant (Thomas, 1991; Oates, 1994).

Sunose has recently (1996) suggested an 'environmental index' to evaluate an environment or site for butterfly assemblages, which ascribes additional value to species which depend on natural conditions. Three major categories of butterflies were recognized: urban (and/or rural) species, 'seminatural' species, and 'natural' species, ascribed values of 1, 2, 3, respectively. The environmental index is the summed values of the species present, and a high value indicates a 'good' environment for butterflies. Trials at several sites in Japan suggested that this index might be of value in monitoring changes in the environment.

Management practices can markedly influence the continuing suitability of any given site for butterflies through the supply of critical resources, and these practices must also often be directed to other ends. The proportion of shade is a critical resource for many woodland butterflies. In the UK, species typical of more open grassland habitats were largely confined to open woody areas with little shade, but a few species prefer shady conditions (Warren, 1985b, 1992, 1993a, 1993b). The amount of shade is influenced by tree height and the width and orientation of rides. Warren therefore recommended several specific steps to manage woodlands for butterflies: a range of rides with varying levels of shade, some very wide (20 metres or more and orientated east–west), others north–south which become shaded more quickly as trees mature; rides of different widths and orientations can produce a range of shade levels, as can rides running between stands of different ages. Traditionally, woodland rides are mown annually along the centre but less frequently (about every five or six years) along the edges. Late summer cutting may be less damaging than early summer cutting, and it is preferable for one small part of a ride to be cut at a time, to provide a mosaic of butterfly habitats.

In a multiple management programme for game birds in southern England, areas of woodland managed for pheasants contained more species and individual butterflies than commercial or unmanaged areas (Robertson *et al.*, 1988). This, again, appeared to be due to the greater width of the rides, facilitating the growth of ground flora on which both pheasants and butterflies can capitalize. Game conservation can be an important or predominant local influence on woodland conservation and management, and so butterflies may benefit from management procedures directed at other biota. This principle needs strenuous investigation and exploitation elsewhere. Although butterflies can themselves be useful 'umbrella taxa' (p. 208) (Launer and Murphy, 1994), they may also benefit greatly from the conservation of other taxa that are regarded generally as being more important.

The White Admiral (*Ladoga camilla*, a species which we noted earlier as having expanded its British range recently) also responds to woodland management practices. Traditional management involved 'coppicing', with trees being cut back intensively at intervals of up to 20 years. Towards the completion of the next cycle there is dense shade, but the larval food plant (Honeysuckle) is often also eliminated because it is regarded as a weed in woodland being managed for timber extraction. Coppice management has been abandoned progressively, leading consistently to conditions more suitable for *L. camilla* (Pollard, 1979). With increasing height of canopy and light penetration, honeysuckle is able to thrive, and this change in traditional management is a significant factor in fostering the range expansion of the butterfly.

In complete contrast, the decline of the Heath Fritillary in the UK has been attributed largely to abandonment of coppicing, which created the continuity of suitable 'open' habitats within small areas of woodland (Warren *et al.*, 1984, Warren, 1987c). This is an example of a species which depends on ephemeral habitats, and management to conserve it must continually provide these conditions (pp. 169–71). If the interruption to habitat supply lasts for around 5–10 years, extinction is probably inevitable (Warren, 1987c). Pollard (1982) clearly showed that different butterflies could respond in markedly different ways to the creation of 'open woodland'. Some species responded slowly, so that they may have responded to changes in food plant abundance rather than to sunlight penetration.

Other habitats may also need critical management to conserve butterflies, and it is sobering to reflect that there are numerous instances where butterflies have become extinct locally in native reserves, and in some cases in reserves designated primarily for their conservation. As Hooper (1971) noted, attempts to conserve rare insects (if these are made at all) on reserves can be on a single reserve with a large and viable population (as long as current conditions prevail), on several smaller sites within the normal range of the species, or in areas where the species may be vulnerable or marginal and where augmentation of populations may be continually needed to ensure its persistence.

Much habitat management is directed towards maintaining earlier successional stages in the usual long-term transition which occurs towards woodland and forest. One consequence of the widespread changes in agricultural practices (for greater 'efficiency') during the last 35 years or so has been that much grassland which was previously lightly cultivated (unfertilized, mown, lightly grassed) is now much more intensively treated or, alternatively, abandoned. As Erhardt (1985) noted, many steep slopes in alpine central Europe — which were difficult to cultivate and unrewarding economically — were especially prone to abandonment and have thus yielded to succession by progressive invasion of shrubs and young trees, with the potential to become climax forest vegetation in due course. Erhardt's data for Switzerland are important in exemplifying what is likely to be a much more widespread suite of trends. His study covered four related topics:

- the effect of abandoning grassland on Lepidoptera;
- the influences of methods of grassland cultivation on Lepidoptera;
- the value of abandoned grassland related to butterfly decline and its significance for conservation; and
- the reaction of the butterfly fauna to vegetational change and correlation between vegetation and butterfly communities.

Diurnal Lepidoptera were assessed by transect counts during the summers of three consecutive years over 35 sites, and species were categorized into one of seven (quantified) abundance categories, from very rare to extremely common. In cultivated grassland there was a high correlation between butterfly diversity and plant diversity. Mowing reduced butterfly populations in several ways, and was more harmful than light grazing and trampling by cattle. It caused uniform breakdown of vegetation structure over the whole area, destroyed flowers utilized by adults and directly killed many of the early growth stages. A comparison of fertilized and unfertilized mown meadows only 300 metres apart showed a strong decline of Lepidoptera in fertilized areas. About 32 species occurred in unfertilized meadows, and 21 (11 butterflies) of these were absent from the fertilized area. Only six butterfly species were found in both areas. Reduced plant diversity in later successional stages led to a decrease in butterfly species, but more still occurred there than in unfertilized mown meadows.

The factors affecting Lepidoptera diversity in these vegetation types are shown in Figure 6.2. As the habitat becomes more and more unsuitable, those species capable of doing so will emigrate, and Erhardt (1985) concluded that butterfly communities act as very sensitive indicators of structure and change — both human-induced and natural — in vegetation.

In the UK, a shift in species emphasis in grasslands from Lycaenidae to Hesperiidae and Satyrinae has been attributed to the cessation of grazing. The general management lesson from a range of studies is that early successional stages needed by particular butterflies must be renewed by rotational cultivation, so that mowing or light grazing are invaluable aids to conservation (Thomas, 1984), although some endangered stages can be preserved only with additional cultivation practices (Erhardt, 1985).

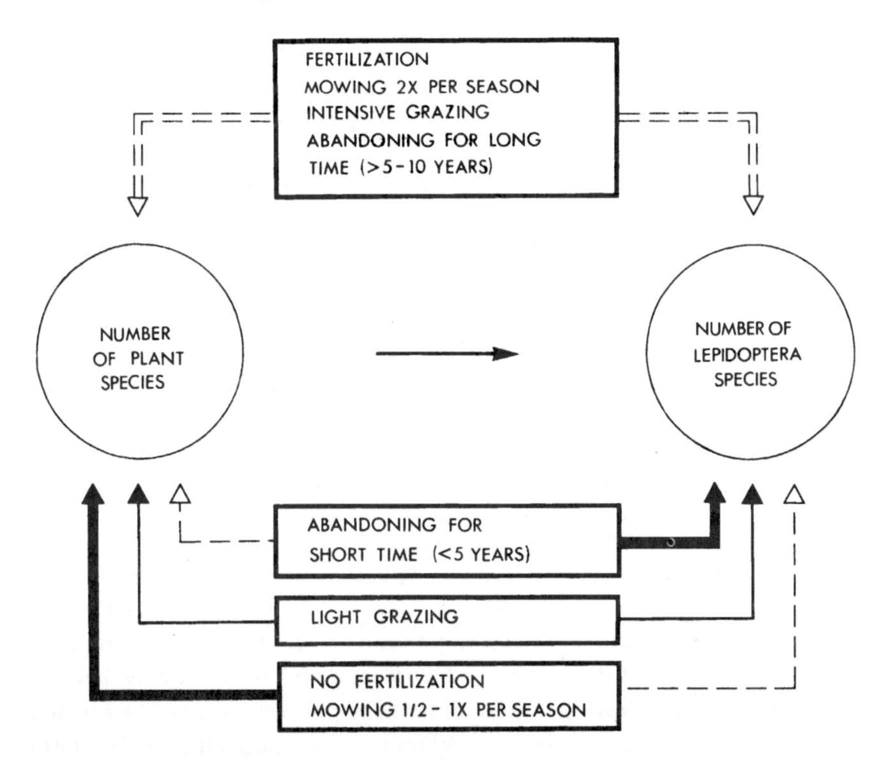

Figure 6.2 Factors affecting plant and Lepidoptera diversity in cultivated and abandoned grasslands. Nature of effect (- -) negative; (= =) strongly negative; (——) positive; (▬▬) strongly positive (Erhardt, 1985).

The Silver-studded Blue (*Plebejus argus*) is one of the most rapidly declining butterflies in the UK, and Thomas (1985) found that it occupied early successional habitats at low altitudes, predominantly on south-facing slopes. Grazing by stock on limestone grasslands has maintained colonies there over long periods, but on heaths the species persists by progressive colonization of newly created pioneer habitats rather than by staying in any one place. Burned areas may provide suitable habitat for around five years, and disturbed areas for 10 years or more. Only large heaths can support *P. argus* indefinitely without deliberate management to provide suitable breeding sites, which need be only 0.5 hectares or so in extent. The decline of the butterfly on heathlands seems to have resulted from the decline of traditional management (furze cutting, peat digging, intentional rotational burning) and progressive isolation and reduction of heath areas (thereby providing barriers to dispersal for the rather sedentary butterflies).

Remedial management for *P. argus* could include maintaining grazing regimes on limestone grasslands and organized burning and disturbance of heathlands. The latter is a principle of much more general application, and one which many conservationists will find alien in face of the widespread dictum of 'keeping areas pristine'. Thomas noted that heathland vegetation which had recolonized bare quarry floors, sandy ground after removal of an aerodrome runway, disused tracks, and areas trampled by tourists have all created suitable areas for *P. argus* to breed. Judicious use of bulldozers to remove mature heath and some topsoil could be a very worthwhile management strategy (see also comments on the Adonis Blue pp. 174–5). Rotational burning requires more study as a management tool for early successional stages but, as noted under 'declines', routine conflagrations through natural burning are part of the normal (or predictable) environmental variation in much of Australia, for example, and the exclusion of fire can permit drastic changes in vegetation. Many plants in such areas are pre-eminently adapted to fire and their seeds may not be liberated or germinate unless burning occurs. Both in Australia and in the prairies of North America, many resident butterflies depend on such fire mosaics for several of their larval and adult food resources. Fire prevention regulations over parts of western North America resulted in the transformation of some coastal grasslands to scrubland or forest, with a consequent disappearance of native butterflies such as *Speyeria* (Hammond and McCorkle, 1984).

In areas where the growing seasons are not continuous, the timing of disturbances such as fire may be critical to the availability of particular stages of a butterfly food plant. For the Black Swallowtail (*Papilio polyxenes*), eggs placed on the leaves of the umbelliferous food plant before flowering suffered nearly three times as much mortality as those deposited on flowers only two to three weeks later (Blau, 1980), although this butterfly is well adapted to colonize local food plant patches produced by disturbances in a continuous growing season.

Butterfly species each have their particular responses to fire (Swengel, 1996), and the influences of burning are thus collectively complex. A number of true prairie species in North America showed negative effects which persisted for at least three to five years, and several such species are candidates for listing under the *Endangered Species Act* (p. 82). In contrast, more generalist invasive species are most abundant in the most recently burned areas and scarcest in areas which have been long unburned. Cutting prairies for hay is more beneficial for the specialist butterflies, and for the butterfly fauna as a whole.

Recently burned sites are acceptable habitats for some rare species, however. *Euphydryas gilletti*, a rare North American nymphalid which inhabits moist montane meadows, is able to exploit newly burned habitats as part of its metapopulation dynamics (Williams, 1995). In Japan, the scarce lycaenid *Shijimiaeoides divinus asonis* prefers burned-out grasslands, because this reduces competition on the food plant. Management by burning may be an important part of conservation for this species (Murata and Nohara, 1993).

The marriage of agricultural practices with nature conservation will continue to cause heated debate. Removal of hedgerows, with their reservoir populations of weeds and other butterfly and caterpillar food resources, has been associated with the decline of some butterflies and also with the progressive isolation of populations by the removal of habitat corridors which could facilitate or encourage movement between sites. Many 'traditional' farmland butterflies in Europe have now been reduced to woodland colonies and are absent from the more open areas which they formerly frequented. Even 'odd corners' of fields left unmanaged may function as reservoir habitats for butterflies. Deliberate management of margins as conservation headlands is now widespread. Roadside verges may be another valuable habitat, and in Europe there has been a welcome trend by councils to reduce the mowing intensity of these areas and even to augment habitat by seeding them with wildflower mixtures, sometimes deliberately biased toward larval food plants and nectar sources for adults.

As Morris and Wells (1987), and many others, have commented, agricultural development has always spelled change for wildlife. Comparisons of the numbers of butterflies on sprayed and unsprayed field edges revealed 13 of the 17 species recorded in more than one transect section to be more numerous on unsprayed edges (Rands and Sotherton, 1986). In total, 868 butterflies were seen on their unsprayed plot and only 297 on the sprayed plot, suggesting that the use of pesticides on cereals may have reduced butterfly numbers on this arable farmland. Rands and Sotherton (1986) suggested that, for bivoltine species in particular, minor modifications in farming practices could increase the value of cereal fields as a butterfly habitat. Some species could complete a generation in unsprayed headlands, and species which disperse little (and which have tended to decline more because of agricultural intensification) may particularly benefit. Recent trends to leave agricultural headlands unsprayed to foster the development of a range of arthropods on which game bird (partridge) chicks depend may well benefit such butterfly species and, again, indicates the direct benefit to butterflies provided by another commercial interest. Although partridge are normally thought of as plant-feeding birds, the chicks need insect (or, at least, arthropod) food during their first few months of life, and spraying field margins, headlands and hedges had led to a decline of partridge in the UK. Conservation of such areas has countered this trend.

The management of broad habitats therefore often affects the butterfly communities present but, as the above few examples emphasize, either decline *or* increased abundance may result. Unless targeting a single species for detailed conservation is attempted in a given area, it may be necessary to evaluate the likelihood of these consequences for a range of different species there. Pollard (1982) used his abundance index from transect counts (pp. 101–8) to attempt to assess fluctuations of this sort, by comparing those from a managed nature reserve with broader trends over the UK. He calculated these broader trends by using 'ratio estimates' of annual change on sites which had been recorded in consecutive years. The 'ratio estimate' is the sum of index values in year 2 divided by those in year 1, so that a value of '1' indicates no change, '>1' an increase and '<1' a decrease. An important result from Pollard's work was the discovery that, for most species, index values tended to fluctuate synchronously over large areas. Local site factors (such as slow habitat change) have only small effects on annual changes but over years will lead to values departing from the broader trends. If a large part of the transect is altered suddenly, the resulting change is likely to override such annual fluctuations. Examples of Pollard's (1982) data are shown in Figure 6.3, for field and ride transects together at Monks Wood; clearly, there is high correspondence between local and regional trends for many species, but particular species have fared either better or worse. Thus, *M. jurtina* seemingly benefited from ride and field management, and *Pyronia tithonus* (the Hedge Brown) benefited from ride management but declined in open field areas. Such results may be difficult to interpret, but indicate priority species which merit further investigation because they are likely to respond to habitat change, and the time scale over which responses may occur. Unfortunately, the opportunity for many studies of similar duration (seven years) to Pollard's are severely limited. As Thomas (1984) wrote, though, land management can 'profoundly alter the carrying capacities of sites for butterflies', and even what appear to be very minor habitat changes may result in changes in birth rate or larval survival.

Again from Thomas, 'All researched (butterfly) extinctions on nature reserves have been due to a failure by conservationists to maintain enough (or any) suitable breeding habitat'. Where resources permit, management to counter this can be attempted, but only once the particular species' needs are appreciated. Some early successional butterflies, for example, may depend on avalanches (in

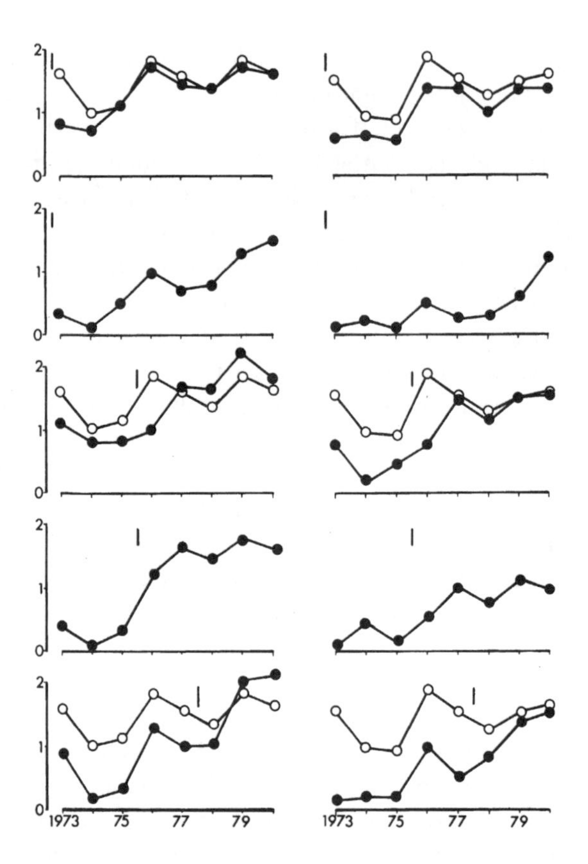

Figure 6.3 Examples of Pollard's (1982) data on the effects of woodland management on butterfly abundance. Index values (log. scale, vertical) for individual managed rides at Monks Wood (five rides, top to bottom of figure) over 1973–80. Solid spots are index values for each year for Meadow Brown (left) and Hedge Brown (right) compared with the summed index values for unmanaged rides (open circles, not shown on all graphs, as line identical for each column); the vertical dash on each graph shows the time of ride widening.

the alps) or cliff falls (*Melitaea cinxia* on the Isle of Wight, p. 105) as modes of natural habitat regeneration. The needs can, of course, be very subtle, especially if the aim of management is to foster diversity, as species co-existing in any habitat can divide up the resource in ways which are often difficult to detect, and any new study tends to reveal different parameters for particular species.

Neotropical Riodininae tend to achieve habitat isolation by mating in different topographic sites, for example (Callaghan, 1983), where their prevailing localities depend not on prominent features, such as hilltops, but on forest micro-habitats based on sun/shadow regimes and time of day. Tropical forest Nymphalidae in Costa Rica (or, at least, the fruit-feeding taxa which can be baited to confirm their presence) are stratified vertically in relation to forest structure (De Vries, 1988b). Most of the 46 species were trapped either only in the understorey or in the canopy. Species richness was similar in both layers, but canopy traps collected more individuals than understorey traps. Members of the Charaxinae and Nymphalinae occurred in the canopy, but Morphinae and Satyrinae in the understorey. Some canopy species could indeed be trapped at ground level at the forest edge because they normally frequented the interface between deep shade and bright sunlight so that, in De Vries' interpretation, they 'treat the forest edge as the canopy come to the ground'.

Wetland butterflies, those frequenting such areas as marshes, bogs and fens, are also commonly restricted to particular micro-habitats (Shuey, 1985). Progressive analysis and survey of any complex habitat tends to reveal similar specializations, and so species lists from apparently similar areas may differ substantially. But, very commonly, butterfly faunas are richest in less disturbed areas and poorer in highly modified ones. In central Spain, the most impoverished of a series of habitats surveyed by Viejo (1986) were irrigated croplands. Dry croplands, in contrast, supported a high diversity of butterflies because of the complexes of marginal plants, such as weeds. Highest numbers in a series of 670 sites examined in a survey in Britain occurred in unimproved pastures (Thomas, 1983c).

Patterns of resource use form one aspect of adaptive strategies of butterflies to their environments. A few attempts have been made to explore these, within the vast context of butterfly biology outlined in Chapter 2. The extent of patchiness of host plants in the tropics, and the patch size, tend to reflect different larval feeding patterns (Young, 1980). For example, forest canopy has a very high 'patchiness' with patch sizes relatively small. These circumstances may favour polyphagy. Secondary successions have low patchiness and large patch sizes, so that species can afford to be monophagous — they can depend, with migration, on obtaining sufficient food from these relatively large stands of suitable plants. Management implications are that the latter habitat may need to be artificially provided and, whereas rare host plants can easily be removed from the forest canopy, many of the butterfly species may be able to feed on other species as long as parts of the canopy are left. This may not be the case for some more diverse groups of insects. I cited Erwin's (1982) Amazonian forest canopy beetles in Chapter 1 — part of the reasoning behind his high estimate of arthropod diversity was that each tree species may have specialist feeders, and the extent of this for some groups of butterflies remains unknown. However, small blocks of tropical forest conserved, for instance, because of traditional 'magical' associations (Larsen *et al.*, 1980: Nigeria) can still support rare forest butterflies. Indeed, one such patch of secondary bush near Lagos supported 385 species of butterflies.

Butterflies, as many other insects, tend to perceive their environments on a much finer scale than people do: what may appear to us to be very similar habitats may be differentiated by butterflies into many different kinds, some suitable and others less so. We have seen that many of the rarest butterflies and those of conservation interest are 'specialist' species. It follows that their optimum habitats and resources may be much more restricted than appears even from a reasonable knowledge of their biology and that management, even when informed, may still be somewhat imprecise in relation to provision of optimum conditions.

CHAPTER SEVEN

CASE HISTORIES IN BUTTERFLY CONSERVATION

This chapter summarizes a number of actual case histories of butterfly conservation. There are many such histories and it is not intended to include more than representative examples here. Any case history is a programme (based loosely on a scheme such as that outlined in Figure 5.1 on page 86) which considers and attempts to conserve a particular butterfly taxon, and yields information which can help in the conservation of other species. Much of the methodology and understanding outlined in earlier chapters has arisen directly from particular cases. Yet each case is ultimately unique, and even the same species in different places may pose different problems and logistic priorities, not least because practical conservation involves the balance of human values which will vary with both space and time. A particular site may assume immense value if, for example, it prevents the establishment of a major highway, shopping complex or tourist resort. Cynics have commented that virtually any proposal for developments of this sort will lead to the discovery of a rare insect, bird or plant there by conservationists. But it is not surprising that many of the best-known examples of butterfly conservation activity result from conflicts over ecosystem use in the most developed parts of the world, while in places where there are no (or few) concerned observers, change may go unheralded to the same extent, or even unnoticed. The converse is that it is only in the well-developed parts of the world that the luxury of expensive autecological studies designed to conserve particular butterfly species or habitats can be supported and publicly condoned.

In general, 'cases' fall into two main categories:

1. Those involving particular populations or rare species threatened by some form of immediate or imminent habitat change, and needing urgent 'crisis management' to conserve them.

2. Those involving study and rectification of more widespread decline over a broader part of their range.

The following examples have been selected to illustrate various facets of both of these. Some have been successful, others not; some are long-term and others are of only short duration or recent incidence. Together, they reveal many of the strategies noted earlier, and demonstrate the attention to detail which characterizes any responsible conservation exercise. However well meant, both factual and judgemental mistakes have occurred, yet as a result the science of butterfly conservation has progressed to being a leading area of invertebrate conservation.

There is some bias in its practice, because particular families of butterflies have aroused much more concern at the 'species level' than others. In particular, Pieridae are very poorly represented as well-documented species-based case histories: although, for example, Heath (1981) noted 15 species of this family as of concern in Europe, few have been studied in any great detail. A similar situation pertains for the skippers. In contrast, many more species of Lycaenidae, Nymphalidae and Papilionidae have been studied, and these must inevitably take precedence in an account intended to exemplify progress and methodology in conservation of particular species. One reason why so many ecologically specialized Lycaenidae have been highlighted is that they may occupy very small areas ('often not larger than a tennis court': Henning and Henning, 1989) so that they may be particularly vulnerable to relatively small disturbances such as the erection of a single building or the ploughing of a single field. Many such instances involving lycaenids were noted in New (1993).

The following cases are alike in having aroused scientific concern. Many have also engendered much concern from the media and the public in reflecting the effects of human intrusion into more natural systems. Many give cause for optimism over the futures of similar cases. The willingness with which many people have given their time and energy to support the wellbeing of local butterflies — sometimes adopting these as local emblems or raising large sums for habitat purchase — has demonstrated clearly that people still have the capacity to care for our world, and will probably continue to do so.

Case histories from many countries have been documented in various Red Data Books (pp. 74–5) and entomological journals, and a recent pioneering book (Hama *et al.*, 1989) discussed a series of 21 species as case histories of butterfly conservation in Japan. Two later volumes appraise additional Japanese species and mark a significant upsurge of interest in the region. Survival plans suggested there for other butterflies are frequently based on, or extrapolated from, these. Most of the following cases are based on single published summary accounts, either in Red Data Books or research journals, and the major references given for each provide much more detailed information. In some cases the situation has changed somewhat since the accounts were published, but they have been selected for inclusion here because of their urgency or differing priorities. Together, they exemplify much of the recent thrust of species-orientated butterfly conservation.

HESPERIIDAE

The Silver-Spotted Skipper, *Hesperia comma*

This skipper was classified as 'vulnerable' in the UK, and its presence is regarded as noteworthy on high-grade wildlife sites. A survey in 1982 (Thomas *et al.*, 1986) showed that it had declined to about 49 UK populations, many of them small. It occurs on calcareous grassland, where larvae feed on a grass (*Festuca ovina*) which is common and widespread on many sites where the butterfly does not occur. The survey also compared conditions in colony-bearing sites with 72 sites where the butterfly had become extinct — some where extinction had occurred decades earlier, so that conditions might have changed substantially, and some where extinction was very recent. Several breeding areas were close together, so that the definition of 'colony' boundaries was sometimes subjective. The density of individuals ranged from 0 to 41 adults/100 m of transect, and the flight areas of colonies ranged from 0.05 to 23.3 hectares. Populations were predominantly closed, with very low rates of emigration.

Females are very particular about where they will lay eggs, and prefer to oviposit on small *F. ovina* plants largely surrounded by bare ground or scree and growing in local sun spots. The ideal habitat was facing south on broken terrain (most sites occupied had a distinct southerly aspect) with a surround of about 45% food plant and about 40% bare ground. Thomas *et al.* believed that the changed status of *H. comma* reflected the extent to which grazing (and, hence, sparse swards) had diminished on the steep land since the late nineteenth century.

Until the dramatic onset of myxomatosis in the mid-1950s, a suitable short sward was maintained on many hills by high rabbit populations. Resultant overgrowing when rabbits were removed coincided with many colony extinctions during the ensuing two decades. Where domestic stock grazing continued over this time, some very large populations survived on a few sites, such as areas where succession is very slow because of skeletal soils.

Since the mid-1970s, rabbits have tended to increase again, and agricultural subsidies have led to more extensive sheep grazing on chalk grasslands. Expansion of range is now occurring, and Thomas *et al.* (1986) pointed out that a number of former sites are now again suitable, although natural colonization of these areas is likely to be slow. They suggest that the number of colonies could be increased to about 75 by introducing adults to uninhabited sites.

It was possible to examine sites in relation to the ideal criteria for oviposition, and Thomas *et al.* found that nearly every site (and sub-site) was well below its potential — the mean condition of occupied sites in 1982 was only 26%, with most of the deficiency being due to insufficient bare ground and almost all the rest to insufficient *F. ovina*. The skipper is clearly extremely sensitive to site condition.

Many *H. comma* sites are on reserves, and the main management needed is continuity of early seral vegetation on slopes facing south, a tactic which can be achieved by heavy grazing — in itself a strategy which may not receive complete sympathy on reserves in which it is necessary (or expected) that many other biota will also be conserved. But if only small areas, of a hectare or so each, can be grazed in rotation, this may do much to enhance the sites for *H. comma*.

Thomas and Jones (1993) suggested that, if the habitat were not fragmented, *H. comma* could be expected to recover its status in south-eastern England in 50–75 years, but because substantial fragmentation has indeed occurred, little further spread may occur because distances of more than 10 km of unsuitable habitat prevent dispersal. Successful conservation requires the protection of metapopulations in networks of habitat patches. Recolonization has occurred largely from refuge areas, but many apparently suitable patches remained uncolonized in 1991 (Thomas and Jones, 1993): the species may be naturally mobile, so that potential habitat patches for the species may be important in conservation planning, in addition to habitats presently occupied. This is likely to be a general consideration in planning the conservation of metapopulation-based species. The broad objectives of the Species Action Plan (Barnett and Warren, 1995c) are to maintain populations at all occupied sites and clusters of sites, restore UK levels to those before the recent decline, and reduce the reliance on rabbit grazing for maintaining habitats in suitable condition.

The Dakota Skipper, *Hesperia dacotae*

This small cryptic skipper now occurs only on remnant native prairie tracts in four States of the USA. Although the population size is unknown (Wells *et al.*, 1983), the univoltine skipper is known from only about 30 sites, many with only small populations which may be largely closed. It is confined to North and South Dakota, Iowa and Minnesota and the habitat is defined as 'mid-continental prairie'. *H. dacotae* is found only where native vegetation is relatively undisturbed, and the major threat has been destruction of habitat for agriculture or ranching, with previously marginal-value tracts gradually becoming incorporated into cropping areas. Grazing by stock has led to a decline of flowers frequented by adults. Some sites are also threatened by gravel mining operations, housing development and irrigation projects. The skipper does occur on several nature reserves but these are small and not properly managed.

One reserve was acquired solely for the Dakota Skipper (Downey, 1981), and the species' biology has been extensively studied, with a workshop held in 1980 to discuss its conservation. Larvae feed on various Poaceae and possibly some sedges (*Carex*), and live in shelters. In the past, wide expanses of habitat would have countered the effects of habitat change, but this is no longer the case (McCabe, 1981). Moderate grazing and occasional fires could have helped to maintain the habitat, and some areas where the skipper is still present have been mowed regularly for hay. Late season mowing and late summer or early spring fires are likely to be the least damaging, as the shelters of young larvae (autumn) are subterranean and post-diapause (spring) larvae are on the ground surface.

It is clearly a prairie specialist (Swengel, 1996) and declined markedly after fire, despite earlier suggestions that larvae might be protected by their sheltering habit (Dana, 1991). Their persistence in low numbers in fire-managed areas may reflect the influences of fire in encouraging a thicker sward, despite a reduction in nectar-rich forbs. The effects of prescribed burning on such species may be very complex.

The habitat of the Dakota Skipper is in need of protection, and the butterfly has been proposed for 'Threatened' status under the US *Endangered Species Act*. More documentation of the populations on reserves is required in order to formulate a management plan. Agencies administering sites where the skipper occurs should be encouraged to consider it in their overall management planning. In contrast to some other species noted in this section, a captive breeding programme was considered undesirable, both because of the difficulties of rearing the skipper in captivity, and because this could impose undesirable selection pressures.

The Altona Skipper (or Yellowish Skipper), *Hesperilla flavescens flavescens*

This endemic south-east Australian skipper has aroused concern for its well-being in recent years, in part as a facet of the general support for butterfly conservation engendered by the Eltham Copper (pp. 182–4). It is associated with open areas, commonly around swamps or lakes where its usual food plant, the sedge *Gahnia filum*, grows. Although distinctive in appearance, the taxonomic status of this skipper is not wholly clear. It is a very light yellow-marked phenotype allied to *H. donnysa*, itself an extraordinarily variable butterfly with several subspecies. Surveys by D. Crosby (pers. comm., 1989; 1990) suggest that it is an extreme clinal form and rather more widespread than earlier supposed. Some colonies of a second subspecies of *H. flavescens*, found in South Australia, have recently become extinct.

Until Crosby's recent survey, the only known substantial colonies were at Altona, on the western outskirts of Melbourne, where they have been long known to collectors in two sites which are under substantial threat from housing and industrial development. Concern by collectors (Crosby, 1986) led to approaches to the land manager, the then Melbourne and Metropolitan Board of Works, for an assessment of its status. The two main Altona populations are still regarded as significant, and possibly constitute an incipient species which does not occur anywhere else except in very small numbers. It is thus of significant evolutionary interest in the development of a group of endemic trapezitine skippers. Members have no potential for interchange with other populations, so that the colonies are essentially closed.

The butterfly probably has two generations each year, but even this is not yet clear. Eggs are laid singly on *G. filum* close to the ground, and the caterpillars make leaf shelters on the sedge. *G. filum* is itself restricted in distribution to unshaded, mildly saline, swampy conditions which are inundated for part of the year, but the latter condition is not necessary for *H. flavescens*. Small plants, with new growth, seem essential for young caterpillars to establish, and oviposition is facilitated if a grass surround does not crowd out the plants. The area of suitable habitat has been more than halved during the last 40 years or so, and protection of the remaining habitat is needed to conserve the subspecies, which is regarded as 'vulnerable'. Numbers at each site are in the order of a few hundred.

The main problems at the two sites are rather different. Site 1 is characterized by the over-maturation of the food plant and lack of regeneration over much of the area (in the past, horse-grazing has destroyed some of the site), weed invasion, localized industrial development on private land next to the main food plant area, and some rubbish dumping through uncontrolled access to the main site via a track used by the public (including trail bikes and some other vehicles). Site 2 is characterized by weed invasion, rubbish dumping and general vulnerability from adjacent housing development. Both sites have suffered from localized burns in recent years and, whereas these have killed some food plants, they have facilitated regeneration of both *Gahnia* and exotic weeds. More severe fires are now a possibility as a consequence of vandalism from the substantially increased local population around site 1, and changed hydrology of the area may have long-term effects.

Conservation measures suggested by Crosby (1990) include:

- controlling public access (including vehicles at site one) to prevent trampling the main colony sites and dumping of rubbish,
- removing rubbish, and controlling major weed pests such as *Phragmites* and thistles,
- establishing a buffer zone against fires, by perimeter mowing between boundary fences and the food plant areas,
- management to ensure regeneration of *Gahnia*, possibly by late winter slashing of small areas on a rotational basis,
- Improving fencing to exclude horses and other grazing animals,
- possibly purchasing of private land abutting the main part of site 1, and

- appointing a ranger to oversee these procedures is also seen as desirable.

There is a slight possibility of over-collecting, because larger larvae and pupae in leaf tunnels can easily be collected systematically. The 'Voluntary Protection Code' to which this skipper is subject in Victoria may help in part to counter this, and there is some local support for its conservation.

PAPILIONIDAE

Queen Alexandra's Birdwing, *Ornithoptera alexandrae*

The world's largest butterfly occurs only in small regions of montane Papua New Guinea, where it is scarce in primary and secondary rainforest. It has 'Endangered' status from the IUCN. It is doubtful whether *O. alexandrae* has ever been common, but nowadays, though legally protected (p. 72), it is highly desired by collectors and is perhaps rarer than ever before because of habitat destruction. It is one of very few butterflies to have been figured on 'wanted' posters circulated by dealers, and is also the emblem on the flag of the Oro (Northern) Province of Papua New Guinea. Formal concern for Queen Alexandra's Birdwing was first evident from its inclusion in the Fauna Protection ordinance in 1966. Since then, much of its prime habitat has been lost to logging and the oil palm industry centred on Popondetta. Parsons (1984) estimated that about 2700 hectares of actual or potential forest habitat is being converted to oil palm, an intensive agro-forestry crop.

The species' restricted distribution is difficult to explain, but until recently it had been recorded only in ten 10 × 10 kilometre squares and is absent from much apparently suitable habitat in this region. However, Mercer's (1992) survey on the Mangalase Plateau implied that *O. alexandrae* might be more abundant there than supposed previously, and Parsons (1996b) commented that it 'is often not uncommon where its larval foodplants occur.' The larvae feed on species of *Aristolochia*, and some authorities have considered them to be monophagous on *A. dielsiana* (formerly treated as *A. schlecteri* in some accounts). However, Parsons (1992c) noted feeding on a group of two or three closely related *Aristolochia*, later named as *Paraaristolochia alexandriana* and *P. meridionaliana* (Parsons, 1996). The vines occur in tall primary forest, the main habitat of *O. alexandrae*, but also thrive in recovering secondary forest, where they are tracked by the butterfly. Around Popondetta, the butterfly occurs in lowland forest up to altitudes of about 800 metres. Two disruptive influences to vegetation in the area have been volcanic disturbance (Mount Lamington, the lower slopes of which support *O. alexandrae*, last erupted in 1951, when it caused substantial local damage: Parsons, 1984) and the development of shifting agriculture on Mount Lamington and the Popondetta Plain. These factors have reduced much of the rainforest to secondary forest. Grasslands, maintained by burning, are also now quite extensive.

O. alexandrae abundance is very difficult to estimate, because it flies very high in the forest and low numbers effectively prohibit any form of MRR study (pp. 95–9). Counts of caterpillars (Straatman, 1970) led to estimates of *maximum* density of 25 per acre. The obvious implication is that large reserves are needed in order to provide even the minimal area needed for a viable population comprising several hundred individuals, but the range of movement of the adults is not known.

The major thrust for the conservation of *O. alexandrae* must be aimed at habitat protection; that is, the maintenance of primary forest areas around Popondetta. Once this is achieved, it may be possible to ranch the species in order to provide specimens for commercial sale to collectors and to augment field populations. This does not appear likely in the near future. It is necessary to obtain as much land as possible for *O. alexandrae*, and some commercial companies are co-operating by planting forested areas on their estates with *Aristolochia* cuttings. Trials on this have been underway for several years; cuttings have been planted at Popondetta in an experimental block with oil palms to determine whether the vine will grow successfully and be exploited by *O. alexandrae* in this habitat.

As a result of a survey by Parsons (1984), three areas where the butterfly occurred in the Kumusi Timber Area were designated as potential *O. alexandrae* reserves. These should be exempted from logging and general access and, eventually, managed for the butterfly. Other small areas not suitable for oil palms may be reservable in perpetuity. A Wildlife Management Area of some 10 000 hectares near Popondetta should also prove a valuable reserve, but the future of this magnificent insect is by no means assured. Recent concerns over continued forestry and agro-forestry operations in the area and the apparent great scarcity of the insect imply that the habitat needs to be protected much more effectively from the effects of human intrusion.

Wildlife Management Areas in Papua New Guinea may be set up as a joint initiative between the Wildlife Division and the traditional land owners, so that there is likely to be strong local support for any such reserves. In the Popondetta region it appears that the Wildlife Division has taken the initiative in attempting to conserve *O. alexandrae*. Concern from biologists overseas, and with little influence in New Guinea, appears at present to exceed that of politicians in the country. It is a perennial problem in conservation matters that, without the commitment and positive goodwill of local people, outside interests tend to be seen as interfering in local affairs and priorities to the detriment of local human wellbeing — there is a very fine line between concern and interference, with consequent alienation of local interests. There are reports of agricultural activity in the Wildlife Management Area (Collins and Morris, 1985). Implementation of the *Conservation Areas Act* (1978), which provides for active management of such areas, is being considered for certain sites, and resolutions for new Wildlife Management Areas are in progress (Collins and Morris, 1985).

Parsons (1992c, 1996b) summarized recent attempts aimed at conserving *O. alexandrae*, and his Action Plan included:

- a core conservation subcomponent designed to minimize the adverse effects of development activities on the butterfly and its habitat;
- a research subcomponent to advance knowledge of the ecological needs of *O. alexandrae* to enhance its conservation management;
- a sustainable development subcomponent to promote long-term conservation of *O. alexandrae* by developing environmentally safe, sustainable, income-generating alternatives to forest felling and similar intrusive practices; and
- an education subcomponent, raising awareness of the butterfly and the broader needs of environmental conservation.

Most of these principles have been independently incorporated into a major international 'Queen Alexandra's Birdwing Butterfly Conservation Project', initiated in 1994 through the Governments of Australia and Papua New Guinea. With funding planned at several million dollars, the butterfly is to be promoted as a major umbrella species to help in the conservation of primary forest habitats by fostering development of rural income from secondary forest areas, in part through the promotion of butterfly ranching. The project, now under way, should add very considerably to our knowledge of this magnificent butterfly and its role in helping to sustain some of the most complex and diverse ecosystems in the region.

The Homerus Swallowtail, *Papilio homerus*

The Homerus Swallowtail, listed by IUCN as 'Endangered' (p. 213), is the largest American swallowtail and is confined to Jamaica. It has several populations, but the two main ones each occupy only a few square kilometres of forest (Collins and Morris, 1985), and the species is restricted to virgin forests on mountain slopes and in gullies. The two populations are probably generally isolated, and the overall abundance seems to be declining. This is an excellent example of a significant species known to be rare and vulnerable and in need of a sound conservation plan.

Decline is due to habitat destruction for both timber extraction and subsequent establishment of softwood and coffee plantations. Commercial collecting by expatriate collectors, although the mountainous terrain renders this difficult, is also considered a likely threat. Some form of safeguarding the remaining habitat seems essential for *P. homerus* to survive, and Collins and Morris (1985) stressed the need for habitat reservation by the Natural Resources

Conservation Department of Jamaica, together with control of collecting. As with many other swallowtails, development of a commercial ranching operation merits investigation, with the aim of satisfying collector demand without depleting wild populations.

Considerable more recent progress with *P. homerus* was reported by Emmel and Garraway (1990), who recommended establishment of patrolled nature reserves or a national park for the remaining areas of habitat, as well as a butterfly farming programme. They emphasized that the plight of *P. homerus* has attracted much attention during the last two decades, and an intensive study of the species' population ecology is now a priority in establishing sound conservation measures. The Xerces Society is currently initiating a major study of *P. homerus*.

During the last few years, *P. homerus* has become an important flagship species for the conservation of upland forest in Jamaica, especially for the Blue Mountains/John Crow National Park areas. According to Larsen (1995c) 'its image is everywhere . . . on tourist literature, posters, wayside billboards and even on pre-paid telephone cards', and it is an important tool in involving young people in conservation.

The Fluminense Swallowtail, *Parides ascanius*

This primitive species, one of a group of brightly coloured swallowtails whose caterpillars feed on *Aristolochia*, now occupies only a small part of its potential range in south-eastern Brazil. Many colonies of *P. ascanius* known before about 1970 are now extinct (Collins and Morris, 1985). In 1973 it was placed on the official list of Brazilian animals threatened with extinction and was for long the only insect accorded that unenviable status (Otero and Brown, 1986), which is intended to protect it from commercial exploitation.

It is a denizen of swampy thickets in dense lowland wooded swamps, and its very inaccessibility tended to keep it safe from collection and habitat development until recently. However, pressure for urban development and recreation has led to coastal and lowland swamps being progressively drained for building, pasture, or banana plantations over virtually the whole range of *P. ascanius*. It is thought that most, if not all, colonies in accessible areas are now known (Otero and Brown, 1986), and the butterfly has become increasingly scarce, especially around Rio de Janeiro.

P. ascanius occurs in several widely separated areas of the 5000-hectare Federal Biological Reserve of Poço dos Antas, where at least 20% of the total area appears to be suitable habitat. This Reserve, established in 1974, has as its main objective the protection of the Golden Lion Marmoset, so that the butterfly could here benefit from more publicized attempts to conserve another charismatic animal in the same area. It may be possible to create new suitable swamps in this reserve.

A small colony also exists in a swampy forest at the Parque Reserve Marapendi, and the habitat there is less than two hectares in extent and is surrounded by alien terrain. A nearby (unreserved) larger colony presumably founded this small one, but the parent colony is threatened with destruction through land subdivision. Otero and Brown (1986) did not regard the reserve colony as viable in the long term. A further reserve proposed at Macae could also play a part in conserving *P. ascanius*. Otero and Brown suggested that the main hope for *P. ascanius* may lie in the inaccessible interior swamps of Rio de Janeiro, which should be safe from any form of development in the foreseeable future. The incidence and distribution of *P. ascanius* there is not yet clear.

Again, commercial ranching on an officially controlled level may be feasible to cater for collector demands, but Otero and Brown noted the high incidence of virus and bacterial disease in the laboratory groups they reared, and suggested that such diseases could be an important cause of mortality in natural populations.

P. ascanius is apparently still quite widespread in its rather small range, but needs to move around over relatively large areas to track *Aristolochia* food plants, for which it competes with other Papilionidae (Brown, 1996). However, because of its undoubted susceptibility to swamp drainage, 'its future remains uncertain' (Tyler *et al.* 1994).

The Apollo, *Parnassius apollo*

P. apollo was the first invertebrate to be listed by CITES (p. 78) in order to monitor and control its trade more effectively. It has long been desired by collectors because of its extensive geographical variation in the Palaearctic, so that a great number of populations have received subspecies or 'varietal' names, and many variant forms are extremely local in occurrence. It is one species for which

Figure 7.1 The Apollo, *Parnassius apollo* (from Kirby, W. F. *Elementary textbook of entomology*. London: Sonnenscheim, 1885.)

over-collecting has been seriously blamed as a cause of decline in many parts of its range. It is thus the subject of numerous local/countrywide legislations and prohibitions in Europe, although such steps do not necessarily safeguard the habitat. Whether or not collecting has contributed to the decline of *P. apollo* is controversial, but it may now be more important than in the past, as many local populations appear to have declined for other reasons (Wells *et al.*, 1983; Collins and Morris, 1985) and are thus likely to be vulnerable. It is 'extinct' in East Germany, 'endangered' in Poland (where a number of populations in national parks have disappeared), and there is concern about several subspecies in France.

Larvae of *P. apollo* feed exclusively on stonecrops (*Sedum*), and major habitat changes which affect the host plants are not tolerated. The greatest threat to *P. apollo* in much of Europe appears to be forestry, especially conifer plantations in relatively large-scale afforestation programmes. Natural habitat succession to scrubland is also a common threat, and various other factors — climatic change (including acid rains), agriculture, tourism, urbanization and others, have also been cited for particular areas. There have been strong calls for habitat regeneration and for the restriction of conifer plantation and scrub succession. It may be feasible in many parts of Europe to restore or recreate habitats, such as in disused limestone quarries in central Europe; this approach was pioneered in Moravia (Kudrna *et al.*, 1994).

An unusual step for augmentation is a suggestion that under artificial conditions it might be possible to rear two generations each year rather than the usual one, so that captive populations could be built up relatively quickly (Nikusch, 1982).

PIERIDAE

The Wood White, *Leptidea sinapis*

This local pierid is, in Britain, confined to parts of the south and west and occurs in discrete and often isolated populations. Its range has become very limited over the last 200 years (Warren, 1981), although there has recently been some colonization of new sites (Warren *et al.*, 1986). Several artificial introductions to new sites have been successful, but colonization potential is otherwise restricted mainly to areas close to existing colonies.

One generation a year is usual over much of the Wood White's British range, but there can be a partial second generation in the south. Larvae feed on a variety of leguminous plants, of which *Lathyrus* and *Lotus* species are probably the most restricted. Adult foraging is also restricted largely to a few perennial herbs, even though many kinds of flower may be available.

As noted earlier (p. 123), decline may have been due to changes in woodland management: in particular, the abandonment of coppicing tended to result in extensive shading of many traditional breeding habitats of *L. sinapis*. Its more recent expansion tends to reflect the converse: the maintenance of wide rides for access and timber extraction, with a central annually mown strip bordered by areas which are cut every five to eight years. Warren (1984) noted that more than half the existing colonies are now on commercial forestry plantations. However, Pollard and Yates (1993) indicated that, as the plantations became older and more shaded, this phase of expansion may have ended, and that, although the Wood White is not in any immediate danger of extinction, its prospects do not seem to be good. Disused railway lines have also proved to be useful habitats. In Britain, several hundred kilometres of lines were abandoned in the 1960s, and many lengths have since developed scrub margins not unlike woodland rides. In some cases, at least, there has been natural spread of Wood Whites to this kind of habitat, which has the potential to form corridors of considerable length.

NYMPHALIDAE

The Monarch, or Wanderer, *Danaus plexippus*

The case of the Monarch is highly unusual, because it is a common and widespread species which cannot be considered in any sense rare; indeed, it is one of the most characteristic large and showy butterflies of North America, and is well known also in Australia and New Zealand (p. 29). Concern over its conservation refers to one aspect of its North American life cycle, namely its spectacular overwintering gatherings in a few restricted sites in California and Mexico, where 'they coalesce by the tens of millions into dense and spectacular aggregations' (Brower and Malcolm, 1989). These aggregations were recognized by Wells *et al.* (1983) as an 'endangered phenomenon' — a spectacular phenological event involving much of the total population, and which is threatened even though the species involved may not otherwise be. Overwintering aggregations are an integral part of the life cycle of this migratory species in North America. Diapausing adults overwinter, mate in spring and migrate northwards and eastwards so that, over a sequence of several generations, much of the continental area is utilized (Figure 7.2). Adults three or four generations removed from the spring migrants return to the same overwintering areas. There are two distinct populations of *D. plexippus* in North America: a western population that overwinters in California and an eastern population that overwinters mainly in Mexico (Nagano and Sakai, 1988). The 'Monarch Project', aiming to conserve the overwintering sites, has been the Xerces Society's largest conservation programme during the last few years, and there is currently a move to designate the Monarch as the United States' national insect, as a symbol of invertebrate conservation with a high profile.

Ten overwintering sites are known in Mexico, and the five major ones are within an area of 500 square kilometres. These are restricted to a few isolated

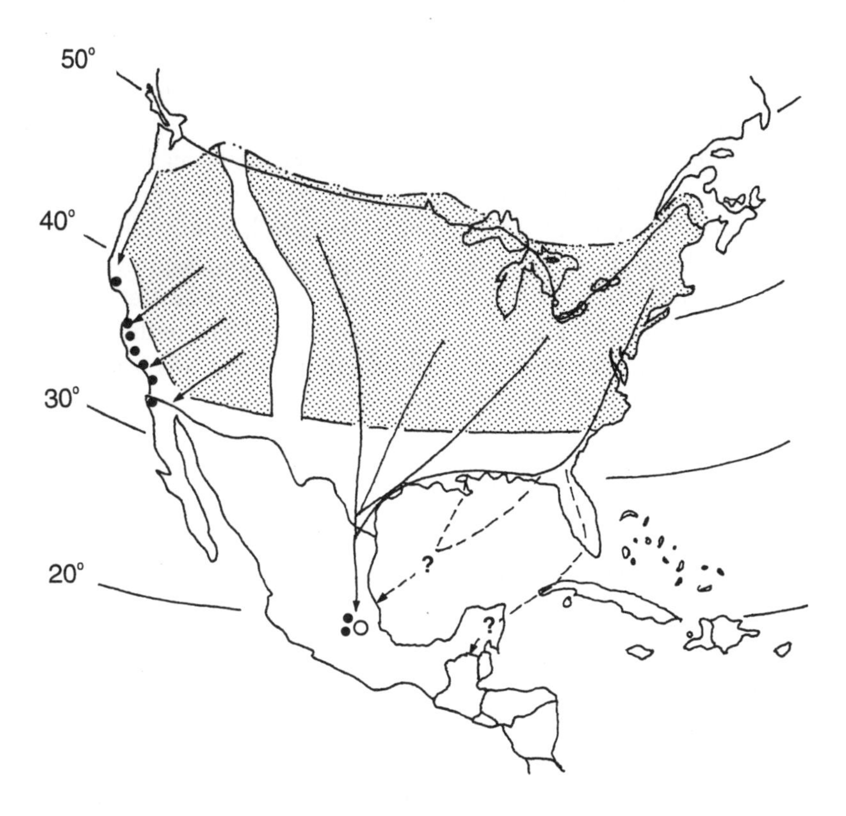

Figure 7.2 Migratory routes of the Monarch, *Danaus plexippus*, in North America, showing the discrete migration paths of western and eastern populations from their extensive summer ranges (shaded). There is a central area (not shaded) of uncertainty, marking the Rocky Mountains (after Brower, 1995).

mountain ranges and to a relatively narrow altitudinal band of around 2900 to 3300 metres, comprising relict Oyamel Fir (*Abies religiosa*) forests, which have been reduced substantially in extent. There are apparently two kinds of overwintering colonies in California (Urquhart *et al.*, 1965): short-duration roosting colonies (which may be abandoned when very severe weather conditions occur) and long-term roosting colonies. Overwintering sites in both countries are threatened, but by rather different pressures, and their conservation is central to the butterfly's conservation (Calvert *et al.*, 1989).

As Brower and Malcolm (1989) stated, forest exploitation in Mexico provides the main or sole means of survival for much of the local human population, and in California habitat destruction reflects the increasing pressures to develop coastal land. There may be about 40 significant overwintering sites in California (Lane, 1984), although many have now been destroyed (Malcolm, 1993). Many comprise groves of introduced *Eucalyptus* trees and virtually all others are widely planted conifers, Monterey Pine (*Pinus radiata*) and Monterey

Cypress (*Cupressus macrocarpa*). The conservation problems of the two regions, both widely recognized, differ substantially but their joint conservation has now been the theme of two major conferences (1981 and 1986), with the latter attracting about 50 participants. No other butterfly has yet attracted this amount of attention for its welfare. *D. plexippus* is the only insect listed under the Bonn Convention, the Convention for the Conservation of Migratory Species of Wild Animals.

The 1986 conference (Malcolm and Zalucki, 1993) ('Moncon-2', following 'Moncon-1' in 1981) brought together most concerned specialists on the butterfly. A more recent meeting in Mexico (Anon., 1993) considered more specifically the conservation of Mexican overwintering sites, now recognized widely as the major key to the species' conservation. Indeed, Brower (1995, 1996) has stressed his belief that unless the oyamel forests are indeed protected from thinning and destruction, the phenomenon of migration and mass overwintering of the eastern population of the Monarch will undoubtedly be destroyed early in the next century.

A Mexican presidential decree in 1986 established an ecological reserve of 16 110 hectares for monarchs, with five conservation centres. Each has a central core (245–1060 hectares) and attendant buffer zones (of up to 6787 hectares). However, tree cutting in these areas appears to be continuing (Brower and Malcolm, 1989), because much of the implementation needed to assure protection is not being carried out. New commercial harvesting of trees and village expansion has led over the last few years to an increased frequency of fires associated with forest clearing. In addition, certain forest pest Lepidoptera have invaded some of the fir areas and these are being sprayed. Although tourist control policies are gradually being developed, the conflicting demands of local, state and federal governments makes this difficult.

Much of the impetus for this Mexican conservation effort has come from an organization known as Monarca AC. Monarca is exploring alternative sources of income for people at present dependent on money from forest exploitation. Some of these are based on tourism, and local people operate as interpreters and guides. Monarca has also commissioned souvenirs such as tiles and postcards to be produced locally. More than 40 000 people visited the overwintering sites in 1985–6, and this number is likely to increase substantially. Tourism is clearly central to the compatability of conservation and the local human population (Ogarrio, 1993).

As in California, the key to successful Monarch overwintering in Mexico is the protection against climatic extremes afforded by the forest. Even moderate forest thinning reduces the forest's capability to retard escape of radiation at night and so the temperature in such areas will inevitably be lower than elsewhere (Calvert *et al.,* 1983). Dew-wetted butterflies are killed at temperatures a few degrees below freezing, but temperatures in the colony do not normally fall this low. Occasional extreme winter storms cause the butterflies to die in large numbers and the role of the dense forest is, essentially, to 'dampen' or buffer the effects of these extremes. The intact forests protect the butterflies by acting as a thermal blanket and reducing the extent to which butterflies are wetted by snow and rain. During the overwintering period, a daily average of 60 000 butterflies per ha land on the ground, with this number increased greatly during storms (Alonso-Mejia *et al.*, 1993). Whilst on the ground they are subject to nocturnal freezing and predation by mice, and the butterflies need to become sufficiently warm to regain the shelter of the trees (Anderson and Brower, 1996). The need to conserve the forest is paramount (Snook, 1993). Brower and Malcolm (1989) also report that disturbance to clustering butterflies which causes excessive flight activity can lead to a high proportion dying of starvation, as their fat reserves are rapidly depleted. Smoke from fires can induce 'frenzied flight' and, in cold conditions, butterflies then drop to the ground and die. A full survey of Monarch habitats has now been completed, but Brower and Malcolm believe that all 10 known Mexican sites should be preserved. It remains to be seen how, and if, this can be achieved. The overwintering populations in California are substantially smaller than those in Mexico, where the largest roosts contain tens of millions of butterflies (Malcolm, 1993).

A major problem with the Californian overwintering sites is that most are on private land and so have little or no guaranteed protection; the fate of the land is at the whim of the particular land owner. Many are threatened with development, and most roosts are within a few kilometres of the coast. Some colonies have been destroyed by urban or agricultural development.

Historically, some of the best-known and best-publicized sites around Pacific Grove ('Butterfly Town, USA') have long been a focus for tourism and nearby development. The town has an ordinance protecting butterflies from molestation, but this does not protect their environment, and development with forest thinning has led to a substantial decline in the overwintering Monarch populations. Brower (1986) noted the sobering (though not particularly unusual, in general conservation terms) fact that, whereas it is against the law to kill one butterfly, it may be perfectly legal to cut down a tree in which thousands of them habitually roost. Lane (1983) likened the situation at such famous Monarch sites to 'killing the goose that laid the golden egg'. In 1986, Californian people passed a land issue to allocate US$2 million for purchasing overwintering habitat, and the Monarch Project is now undertaking the detailed research needed to set priorities for this.

One innovation of the Monarch Project is the 'conservation easement' approach, whereby a land owner may transfer the rights to those parts of the area which support Monarchs on his or her private property, thus giving the Monarch Project possible legal recourse if the habitat is degraded (Allen and Snow, 1993).

The impetus for Monarch conservation on its overwintering areas has flowed to other areas where the butterfly does not roost during winter, but where its breeding habitat may be threatened. The species was first noted in Bermuda in 1859 (Hilburn, 1989), and it does not migrate or diapause there. The Bermuda Department of Agriculture and Fisheries instigated a two-part Monarch Conservation Project in 1988, as many open areas where the food plants formerly grew are now developed for houses and lawns. An educational thrust is promoting the growing of two species of *Asclepias* (milkweeds), *A. curassavica* and *A. phycocarpa*, as ornamental plants (both of these larval food plants have very restricted distributions in the Islands) and a butterfly trapping programme was implemented to gather information on movement and population sizes. The Bermuda climate eliminates the need for Monarchs to migrate and, although individuals arrive there from the north and west in September and October in most years (Hilburn, 1989), numbers are generally small.

The Bay Checkerspot, *Euphydryas editha bayensis*

About 200 years ago this might have been one of the most abundant and widespread butterflies of the California East Range, yet its fate now rests on the persistence of a single population (Murphy and Weiss, 1988). It is also one of the most intensively studied of all butterflies, having been the subject of a sequence of research projects continuously since 1959.

Considerations of the dynamics of the Bay Checkerspot have been important in developing an appreciation of metapopulation dynamics (p. 33) (Murphy *et al.*, 1990), in which the sensitivity to differences in rainfall and insolation of different slope aspects rendered local populations variously susceptible to extinction (Figure 5.5, p. 107).

It is now restricted to grasslands on outcrops of serpentine soils where the larval food plants (*Plantago erecta* as the main one, two species of *Orthocarpus* as secondary hosts) and adult nectar sources occur. This habitat now occurs as small remnant patches, with fragmentation occurring initially when grassland areas other than those on serpentine soil were replaced by introduced grasses and forbs. In the early 1950s around a dozen populations of *E.e. bayensis* still thrived, but most of these have now been destroyed by a variety of processes (Table 7.1). Several colonies in the Santa Clara area went extinct in 1977 but have apparently been recently recolonized. Habitat quality is clearly the central theme for conservation of *E.e. bayensis*, and Murphy and Weiss (1988) suggested that prescriptions for conservation must concentrate on habitat preservation. Specifically:

(a) The sole large population (Morgan Hill, Santa Clara) must be preserved, as it is a likely reservoir population from which natural or artificial colonization of other areas may occur.

Table 7.1 Causes of previous extinctions and current threats to populations of the Bay Checkerspot butterfly, *Euphydryas editha bayensis* (after Murphy and Weiss, 1988).

Population	*County*	*Causes of Extinction*	*Current Threats*
Twin Peaks	San Francisco	Urbanization	
San Bruno Mountain	San Mateo	Invasion by non-native plants, fire	
Hillsborough	San Mateo	Freeway construction, suburbanization	
San Mateo	San Mateo	Freeway construction	
Edgewood	San Mateo		Golf course development
Woodside	San Mateo	Freeway construction, suburbanization	
Jasper Ridge	San Mateo		Climatic fluctuations
Oakland Hills	Alameda	Urbanization	
Franklin Canyon	Contra Costa	Unknown	
Morgan Territory	Contra Costa	Drought, overgrazing	
Silver Creek	Santa Clara	Drought, overgrazing	Suburbanization
Morgan Hill	Santa Clara		Resource development, extreme drought
Coyote Reservoir	Santa Clara	Drought, overgrazing	
Hale Avenue	Santa Clara	Drought	

(b) The few remaining habitat patches in San Mateo must be fully protected, and no further disturbance allowed. The San Bruno Mountain population, now extinct (and unusual as the only one not on serpentine soils) should be re-established using Morgan Hill stock.

(c) The Jasper Ridge and Edgewood Park colonies may well become extinct. In Jasper Ridge, one small population had an effective population size of no more than 15 individuals for four years (Mueller *et al.*, 1985). In the event of extinction, reintroduction from Morgan Hill stock may be feasible.

The political history of conservation concern for the Bay Checkerspot was also summarized by Murphy and Weiss (1988). It was petitioned as a candidate for the USFWS Federal Endangered Species List in 1980, but the listing proposal did not appear until 1985, as listings virtually ceased from 1981 onwards.

The Morgan Hill colony was discovered only in 1983, as the land was until then inaccessible private property. A habitat conservation plan compatible with solid waste landfill deposition was initiated by a commercial company. This plan, in which the company co-operated with the USFWS, included:

- setting aside the best quality butterfly habitat close to the landfill site;
- providing funding for further studies of the butterfly's ecology and genetics;
- regulating grazing activity to encourage high densities of larval food plants and adult nectar resources;
- providing funding for identifying suitable empty habitat patches and re-establishing the butterfly in them.

In contrast, a defence contractor who owned neighbouring parts of the habitat challenged the listing proposal and attempted to prove that the butterfly was, in fact, more widely distributed than earlier documented, and that the overall distribution had not been portrayed accurately by the petitioners.

After considerable debate on taxonomic status, in which the assessment panel adopted the stance that traditional acceptance is a sound criterion for assessing the validity of a subspecies and that taxonomic disagreements motivated by self-interest are not legitimate challenges, *E.e. bayensis* was listed as a 'threatened species' in September 1987 — seven years after the initial petition was filed. Murphy and Weiss point out that it is now the only animal which is now protected and totally restricted to native California grasslands, so that it is an 'umbrella species' for the protection of the remnant communities in which it occurs. Indeed, *E.e. bayensis* is one of very few species whose 'umbrella capacity' has been evaluated specifically (p. 209) (Launer and Murphy, 1994).

The Uncompahgre Fritillary, *Boloria acrocnema*

This fritillary was first discovered only in 1978 in an alpine meadow on Mount Uncompahgre, south-western Colorado (Gall and Sperling, 1980). It aroused interest, not only because of its extremely narrow geographical range, but because it was the first new and previously uncollected butterfly from the mainland USA for nearly 20 years. At its type locality, *B. acrocnema* was restricted to only about one hectare. A second colony found in 1982, also in the San Juan Mountains, is likewise small. It appears to be a glacial relict species, and occurs only above 4000 metres (Gall, 1984a,b). The species declined substantially during the 1980s, apparently to extinction of the type colony. It was listed as 'endangered' by the US Fish and Wildlife Service in 1991 (Opler, 1991).

The only known larval food plant is a prostrate dwarf willow, *Salix nivalis*, but adults take nectar from a variety of alpine flora and bask in direct sunlight. Gall (1984b) studied the population at Mount Uncompahgre by MRR methods on 15 subsites, and (in addition to an ink number to identify each specimen, p. 95) he marked a forewing with bright red to render the specimens less attractive to commercial collectors, who were already (1980) perceived as a threat to the species. Specimens have been advertised at more than US$100 each, and at the time of Gall's account there were no legal restrictions on collection and commercial sale.

Both sexes are sedentary, with mean dispersal ranges of only 50–60 metres, although females tend to move further as they get older. Both known colonies are located in dense stands of *S. nivalis*, and only a few hundred, almost certainly less than 1000, adults occurred each year in the original colony. None was found there in 1987 and 1991, but it has since been found again in small numbers (Britten *et al.*, 1994).

The position of the colony's residence at Mount Uncompahgre moved some 75 metres over a period of about three years. This is a rather unusual feature in butterflies and may be related to the females' age-variable dispersal. Gall (1984a) suggested that females are likely to lay eggs on the site where they initially occur but, having thus 'stabilized' the population's incidence to some extent, may later successfully colonize other patches of *S. nivalis*. Continued survival could therefore reflect a balance between local extinctions and parallel colonizations of other areas. For each site, persistence could reflect a continuous sequence of these events. *If* this occurs, it is clearly very relevant when considering conservation measures, not least because the tactic of translocating females from Mount Uncompahgre to other apparently suitable sites could closely approximate a natural process rather than be wholly artificial; this is in marked contrast to many other cases where translocations have been considered. Gall speculated that the real buffer against extinction of *B. acrocnema* is this pattern of continued mobility and colonization within a patchy habitat.

The type locality has been designated (in December 1980) as a Natural Area by the Colorado State National Area Council, and searches are undertaken regularly for other colonies in the San Juan Ranges. Gall (1984a) considered that sheep grazing (prohibited at the Mount Uncompahgre site) represents the most serious threat to *B. acrocnema* colonies and that a hiking trail bisecting that colony apparently poses little threat. Policing of the sites to control collecting, in conjunction with legal protection, is also needed.

The recent decline may have been due in part to persistent drought conditions. But despite collecting prohibitions, Britten *et al.* (1994) observed collectors at both main colony sites in both years of their study, and believed that they could be implicated in the decline. These authors endorsed Gall's earlier suggestion of policing the colonies; because the flight season lasts only about three weeks, this may not be prohibitively expensive. The major management options available in the mid-1990s were seen as:

- protection of the main colony from human disturbance, particularly from collecting, mineral exploration and livestock grazing;
- careful, non-intrusive monitoring of the population and habitat;
- protection of sites where *B. acrocnema* has been seen previously (including the third site found in 1988, where a single individual, only, was seen some 50 km from the other main colonies: Britten *et al.*, 1994), and visiting them several times each flight reason to determine whether the butterfly persists; and
- diligent searching for new colonies.

Britten *et al.* (1994) also considered that the low levels of genetic variability in the Uncompahgre Fritillary may hasten extinction, through high levels of inbreeding increasing vulnerability to chance catastrophic events, and believed that it was indeed close to extinction. However, some resurgence was evident after their paper was written, and this has not yet been assessed in detail.

The Ptunarra Brown, *Oreixenica ptunarra*

This rare Tasmanian endemic satyrine is highly characteristic of *Poa*-dominated native grassland in the centre of the island state. The species had declined substantially by the 1980s, mainly due to loss of habitat, and was listed as endangered (Prince, 1988). A three-year project to investigate its conservation status and needs was initiated from 1990, and formed the basis for a management programme (Neyland, 1992, 1993).

The flight season of this univoltine butterfly lasts only about two weeks at any site, with the exact period reflecting weather conditions during development, so that adult surveys are necessarily concentrated. In a broad survey over much of the range of all three subspecies of *O. ptunarra*, Neyland located about 150 colonies, many of them previously undocumented; however, over the same period, four of the 33 colonies noted by Prince became extinct. *All* colonies were regarded as threatened to some extent, from the effects of grazing and clearing: whereas it can tolerate some light grazing, *O. ptunarra* soon disappears from more heavily stocked areas. The total area of habitat, estimated at 4000 hectares by Neyland (1993), is continuing to decline. Most extant colonies occupied sites of less than 10 hectares, and only 20 exceeded 50 hectares.

The major requirement for the butterfly was healthy areas of *Poa* tussock extending over at least one hectare. In the major grassland areas, all colonies were subject to grazing, with only one site considered secure. Elsewhere, conversion to pasture and to eucalypt plantations by private timber companies threatens most other colonies, and all three subspecies are considered at risk. Repeated burning of grasslands has also caused a substantial population decline in some areas. In general, suitable *Poa* grassland has declined massively in Tasmania, together with the other inhabitants of this specialized environment.

O. ptunarra is an important umbrella species in this habitat. Measures specified for a recovery plan include the need to fence colonies or otherwise reduce grazing intensity, as a key influence on the butterfly's survival. Neyland recommended that as many 'core areas' as possible should be fenced, that the frequency of burning be reduced, and that wildlife habitat be conserved within plantation development. Long-term monitoring is needed to clarify more fully the effects of grazing on the butterfly and to quantify its fluctuations. Lastly, reintroduction to some former sites may be practicable if these can be adequately restored and safeguarded.

The Heath Fritillary, *Mellicta athalia*

Recent management of sites of this British fritillary appears to have been successful at the last possible moment, at a time when it was on the verge of extinction in the country, having declined steadily for more than 70 years. It occurs in open woodland, and by 1984 Warren *et al.* regarded it as 'probably the most endangered resident butterfly in Britain'. It was then restricted to five sites in south-west England and to three blocks of woodland (totalling 31 colonies) in south-east England, reflecting the remnants of a formerly broader (but still bisected) distribution range.

M. athalia occurs mainly in open sunny habitats at an early stage of woodland succession or regeneration. Eggs are laid in batches and caterpillars feed communally on *Plantago lanceolata, Melampyrum pratense*, and perhaps other plants. Many of the colonies noted by Warren *et al.* (1984) were considered vulnerable because they contained fewer than 200 adults at the peak time and occupied areas of under two hectares.

Causes of decline were for a long time unclear, and included accusations of over-collecting. Warren (1987a,c) showed that colonies were indeed vulnerable (15 Kentish colonies known in 1980 had died out by 1984, but 10 new ones had established), and believed that changes in coppicing practices and clearance for conifer planting were also involved. Colonization of nearby cleared or coppiced habitats sometimes occurred the first summer after cutting, and habitats are unsuitable only five to six years after that. Cessation of coppicing has led directly to decline of habitat. Colonies in vigorous coppice become extinct by the fifth year after cutting, and in conifer plantations by the ninth year. Extinctions may be countered locally if the supply of new habitat is within the colonization range of an existing population (about 300 metres for rapid colonization, or up to about 600 metres for slow colonization: Warren (1987b). No, or very little, colonization is likely to occur from distances of more than a kilometre. One management tactic might be to reduce the effective distance between source colonies and probable new sites in large woodland complexes by providing interconnecting open rides and creating numerous accessible new receiving sites for dispersing butterflies. Many new reserves have been established during the 1980s for the butterfly, and management is thus occurring. Deliberate introductions are needed for widely separated sites, and this tactic appears to have been successful already on a site some 60 kilometres from the nearest other colony.

A large increase in population size on one national nature reserve has been observed since intensive conservation management was investigated. New areas of one to three hectares were cut or coppiced each year from 1979 onwards, and the population, estimated at less than 20 adults in 1980, rose to about 600 in 1984. Warren (1987c) attributed this increase directly to enhanced management. A total emergence of around 3000 adults occurred in 1985.

The successful conservation programme for the Heath Fritillary was summarized by Warren (1991). By 1989, about 80% of colonies were protected to some extent, with some owners positively managing the habitat for the butterfly. Warren's earlier (1987c) recommendations had then been in place for several years on some sites, with success recorded in examples of all the three major habitats of unimproved grassland, sheltered heathland, and coppiced woodland.

The Heath Fritillary exemplifies well the total dependence of some butterfly species on anthropogenic habitats (Thomas, 1991).

The Giant Purple, *Sasakia charonda*

This spectacular nymphalid has been designated as the national butterfly of Japan, but is 'threatened everywhere because of the decrease and fragmentation of large coppices' (Ishii, 1996). Thus, a population in Nagano Prefecture was threatened by deforestation undertaken for a new road in 1990–91 (Hirukawa, 1993). Caterpillars feed on the foliage of Hackberry (*Celtis sinesis*, Ulmaceae) and adults feed mainly on the sap of *Quercus*, so that the species has been considered a good indicator for the condition of deciduous woodlands.

Concern over the decline of the species as a result of land development has exploited its value as a 'symbol' for quality woodland, and led to the formation of a committee being formed in 1979 to protect it (Bandai, 1996). The number of hibernating larvae each winter has been measured since 1980 in Nagasaka-cho. This procedure is straightforward, as larvae pass the winter in leaf litter on the ground near the tree. Counts in November and April give estimates of hibernation success. The Committee was also instrumental in establishing a 7.5 km nature trail near Nagasaka-cho, and this is used as the route along which adult butterflies are counted in summer. New habitat for *Sasakia* is being established by planting new food plants, with the aid of school children, as has occurred also in Saitama Prefecture (Makibayashi, 1996).

As the Giant Purple (known also as the Great Purple Emperor) is one of the town symbols of Nagasaka-cho (the other being clean water), an annual 'Giant Purple Week' festival is a major vacation event for local people, and the butterfly is thus an important symbol for nature conservation and education. This example demonstrates well the impetus for conservation that can be gained by support of the local people (as evident also in some of the other examples outlined in this Chapter), and this resource is indeed vital when major construction or urban development is seen as a major threat.

LYCAENIDAE

The Large Blue, *Maculinea arion*

The Large Blue became extinct in the UK in 1979, and is of particular interest as a species for which there had been active interest in, and active management for, its conservation for many years before this occurred. This interest was especially intense during the final decade or so of its existence. The likelihood of its extinction in the UK was recognized many years ago; it was thought to be imminent in the 1880s, for instance. *M. arion* had always been local in southern England, but occurred at about 30 sites in the early 1950s and from then onwards its history is one of progressive isolation through alienation of habitats leading to a mosaic of small colonies. About half the colonies were destroyed by ploughing, quarrying, urbanization and afforestation (Muggleton and Benham, 1975), and the disappearance of the larval resources after the removal of grazing stock also appeared to be implicated. Some sites apparently changed little, but the butterfly still disappeared and this could not generally be attributed to collector activity. As a result of all this, surviving colonies became very isolated, with genetic isolation possibly becoming a factor leading to further decline.

M. arion depends not only on a specific larval food plant, Wild Thyme (*Thymus drucei praecox*), but also on the presence of *Myrmica* ants, as the caterpillars overwinter and pupate in the ant nests. The caterpillars feed on thyme when young, and older larvae are carried into their nests by the ants and feed on ant larvae for about nine months. The predominant ant involved is *M. sabuleti*, which is very sensitive to grazing regimes on the open ground in which nests are made. *M. sabuleti* is confined to very closely grazed sward; if grazing is stopped, most nests disappear quickly. Grazing also assists the food plant by removing or cropping grasses which would otherwise crowd out the thyme. From about the early 1950s, the grazing of grasslands was progressively eliminated — sometimes deliberately as a conservation measure, but also reflecting the heavy toll of myxomatosis on rabbit populations and changing agricultural practices. The ants declined in many Large Blue sites once the vegetation persisted at more than about five centimetres in height.

In retrospect, some well-intentioned management practices can now be seen to have been misguided, however laudable and forward-looking they might have appeared at the time. They were co-ordinated from 1962 by a 'Large Blue Committee' comprising members from several major conservation bodies.

One of the best sites had been declared a nature reserve in the 1920s, when collecting was thought to be a prime cause of extinction (Thomas, 1977, 1984). The site was fenced, but this also excluded grazing animals, so that the area became overgrown and the butterfly disappeared. In this case, as Thomas has emphasized, informed reserve management came just too late to save the species in the UK. It became easy to modify the habitat to increase nest densities of *M. sabuleti* (Thomas, 1980) but before this the combination of larval overcrowding in ant nests and drought on the last colony reduced it to only five individuals. Crowding was manifest — up to 40 larvae entered a single ant nest, which normally supports only one caterpillar — and crowding resulted in all the larvae dying. The five adults had been reared in semi-captivity as a 'last-ditch' survival attempt in 1978. One female mated, and 59 caterpillars were introduced into ant nests. Twenty-two butterflies emerged in 1979, but these failed to mate.

This case is indeed sobering, and stresses yet again the need for an intimate knowledge of any particular butterfly before it can be managed adequately for conservation. Long-term concern, including habitat 'protection' against perceived threatening processes, proved futile. Intensive study of *M. arion* only a few years earlier could almost certainly have prevented extinction. But, as Thomas (1980) showed, most of the measures which *were* taken were themselves pioneering, as there was no precedent for attempting to conserve any similar myrmecophilous lycaenid.

Suggestions of reintroducing the Large Blue to the UK were made from the 1970s onwards, at a time when it was still present on three or four sites. Since 1982, research and trial introductions of *Maculinea* from Europe have been planned in the expectation of reinstating this widely appreciated butterfly in parts of its former UK range.

The first such introduction was made in 1983, and the population increased for six years, then fell sharply during two years of extreme drought (Thomas, 1994). Other, more recent, introductions to other sites are also encouraging (Thomas, 1994, 1996). One problem, in view of the endangered status of *M. arion* over most of its European range, was finding sufficiently secure donor populations to provide initial stock for reintroduction; a colony from Sweden was eventually selected.

The vast amount of work undertaken on the biology of *M. arion* and its congenors has broken much new ground in advancing butterfly conservation science. Much of this was summarized by Feltwell (1995). It demonstrated clearly the need for detailed understanding of the ecology and interrelationships of the butterfly, host ant, food plant, and habitat management, as well as drawing attention for the first time to the needs of conserving specific ichneumonid parasites as part of the species complex (Thomas and Elmes, 1992; Thomas, 1994). Thomas (1994) noted the value of this case in promoting the fundamental worth of basic ecological understanding, which led directly to funding for other UK butterflies such as the Heath Fritillary (p. 169). The complexity of needs of *M. arion* larvae, and those of related *Maculinea* species, enabled the definition of some habitat requirements for conservation (Elmes and Thomas, 1992) as follows:

- There must be enough food plants producing the right stage of flowerbuds at the time of butterfly oviposition.
- There must be sufficient nests of the correct *Myrmica* species for many food plants to be within foraging range (about 2 m) of a nest.
- The food plants should be dispersed, because if they are too aggregated a few nests accumulate all the *Maculinea*. This leads to competition for ant larvae as food and eventual mortality through exhaustion of food.
- The ant nests must be sufficiently large (about 400 workers) to produce the 230 larvae needed as food by a single *M. arion* larva.

The Adonis Blue, *Lysandra bellargus*

This species has always been locally distributed in the UK, and colonies are confined to chalk grassland. Caterpillars feed on a vetch, *Hippocrepis comosa*, and are tended by ants. During recent years *L. bellargus* has become much rarer, and in 1983 Thomas estimated that only 70–80 populations survived, most of them small, and that the number of colonies had halved every 12 years since the early 1950s. Habitat change was again the major cause of this, with agricultural conversion probably the major factor involved. Many of the colonies which still existed in the early 1980s were considered vulnerable to this trend and to changes in stock grazing regimes.

The colonies are very sedentary, and large populations of food plant typify most sites. Because *L. bellargus* is on the northern fringe of its range in southern England, the heat requirements of early stages may also be critical, and these are affected markedly by the height of the grass sward. Thomas (1983b) showed that a viable colony needs at least 1.2 hectares of grassland, on which the *Hippocrepis* must be present at densities greater than 0.2 plants per square metre. Warmer sites, those on slopes facing south, were favoured, and the removal of grazing led to the substantial decline of some colonies. Females lay eggs almost solely on *Hippocrepis* growing in very short turf; unimproved grasslands taller than five centimetres were not suitable for the butterfly.

Thomas (1983b) made some constructive suggestions for the conservation of *L. bellargus*:

- Reservation or protection of habitats supporting colonies, and management to be commenced rapidly once this occurs. Several colonies have become extinct on nature reserves because of lack of management.
- Rotational management regimes are needed so that early seral stages are maintained and grassland are kept low. Preference must be given to slopes facing south rather than just 'anywhere' on a site.
- Habitats must not be fragmented, even by valleys or patches of scrub, which can constitute effective barriers for the butterflies.

Thomas also raised the controversial point of habitat creation by 'unnatural' means (p. 143). The incidence of south-facing sunny spots on uneven terrain so attractive to the Adonis Blue could, for example, be increased by using heavy earth-moving equipment or explosives! Because *L. bellargus* adults do not move around much there is little possibility of natural colonization of new sites, but reintroduction to sites which were formerly occupied and which are now managed to provide suitable habitats may be a viable conservation option.

The species was introduced to a National Nature Reserve in 1981, where the population increased to around 5000 by 1984, but thereafter crashed during the poor summer. It became extinct there in 1989, and Oates (1994) suggested that the need for hot microclimates rendered the grazing regime, by light sheep, insufficient. Perhaps grazing by heavier animals, such as cattle or a heavier sheep breed, might have provided environments such as hoof-prints which might have helped save the colony.

The species exemplifies well the importance of microclimate to species on the extreme fringes of their natural range. Indeed, Thomas (1996) noted that such species may really be living *beyond* their natural range, where thermal requirements have been maintained only by traditional forms of land use for the last several thousand years. *L. bellargus* is one of several species which has received much attention in the UK but is relatively secure and widespread further south in Europe.

The Large Copper, *Lycaena dispar*

The endemic UK subspecies of this spectacular lycaenid once occurred in a number of wet fenland habitats in eastern England, but rapidly declined in the 1840s. The last specimen of *L.d. dispar* was captured in 1851 (Duffey, 1968, 1977). Fenland drainage undoubtedly led to its decline, as habitat was lost, but this was probably a case where over-collecting of a highly desirable species hastened its extinction in a declining habitat. But, as with *L. bellargus, L. dispar* in the UK was on the fringe of its natural distribution and may thus have had only a very narrow range of optimum ecological conditions under which it could thrive. There was a profitable local trade in the early nineteenth century, as fenmen collected larvae and pupae of *L.d. dispar* for sale to collectors.

Early attempts at re-establishment involved the more widespread European relative, *L.d. rutilus*, and failed, but a programme involving the Dutch subspecies (*L.d. batavus*) has continued from the mid-1920s. Woodwalton Fen National Nature Reserve, the focal site of these introductions, has been managed continuously for *L. dispar* since 1926, when nearly nine hectares were cleared of bushes and planted with the larval food plant, Great Water Dock (*Rumex hydrolapathum*), and butterflies (38 individuals, comprising 25 males and 13 females: Duffey, 1968) were introduced in 1927. More than 1000 butterflies were seen in 1928 and — with sporadic augmentation from captive-bred stock at times when the population was low — the colony persisted until 1969 with this form of management, which led to a great deal of fundamental knowledge about the species.

An unusually severe flood in July 1968 submerged the food plants, so that oviposition could not easily occur. Very few eggs were laid and only five adults (one male, four females) resulted in cages in the nature reserve. No eggs were found subsequently. No caterpillars were found in spring 1970, after their normal period of hibernation on old fallen leaves on the ground.

L.d. batavus was again introduced in 1970, from captive stock. More than 500 individuals of each sex were released, and releases were repeated over the next three years. Mortality from oviposition to spring emergence in each year from 1970–73 was well over 90 per cent, mostly as eggs and the two instars before hibernation. Larval survival is markedly enhanced if they are reared in muslin cages in the fen.

At Woodwalton, then, it seemed that augmentation or protection of the field population may be needed to ensure its persistence. Only about 30 hectares can be maintained in a condition suitable for *L. dispar* to breed, and Duffey (1977) identified four factors which may be needed to counter this habitat restriction: protection of some of the larvae (muslin cages in the field in spring to reduce levels of attack from predators and parasites), creation of germination conditions for *Rumex*, a controlled grazing regime, and maintenance of a captive stock for augmentation or reintroduction, if necessary.

The condition of the water dock plants is critical for the butterfly, and changes in traditional fenland management affected this adversely. Avoidance of plants along the water's edge by female butterflies had long been known. These plants are unusually vigorous and tall and it seems that they are not desirable for oviposition. Females prefer more distant and generally declining plants away from the water. Cattle grazing during June and July renders these more accessible to the butterflies — both by direct grazing and by trampling of surrounding vegetation — and selection of plants away from water is also likely to reduce the risk of caterpillars drowning in floods. In its native Netherlands, *L.d. batavus* was apparently maintained naturally by the traditional practice of peat-digging, which ensured the establishment of early stages of fen succession with continuing germination of water dock. Later exploitation of fens for reeds and hay tended to involve destruction of *Rumex*, and large numbers of food plants are necessary to maintain a viable population of Large Coppers. Continued renewal of plants is clearly necessary to provide a supply of suboptimum or declining plants on which the females will lay. Traditional peat diggings can be imitated by making an annual series of shallow cuttings which are sufficiently wet for *Rumex* seeds to germinate and survive. Regeneration of the dock ceases as succession proceeds.

Considerable help, and therefore expense, will be needed to ensure that *L. dispar* persists at Woodwalton Fen. Cattle grazing is relatively cheap as a management tool, but continuous captive rearing is much more expensive. Heath *et al.* (1984) noted that *L. dispar* had persisted for several years without help from 1979, but was augmented from captive stock from 1971–73 and at intervals subsequently, reflecting large annual reductions in population size.

A broader approach to the restoration of this butterfly in the UK was initiated in 1987 (Pullin *et al.*, 1995), following the practical realization that maintaining a self-sustaining population at Woodwalton Fen was impracticable, either because of the small size of the habitat or an insufficient understanding of the species' requirements. The recent Action Plan (Barnett and Warren,

1995d) embodies the broader objectives of determining the feasibility of re-establishing the Large Copper in the UK and, if so, the possible restoration of a network of populations within the framework of good fen management.

From 1987, a large release of adults was monitored annually, but it declined rapidly to 1989, continuing to almost certain extinction in 1994. Following additional studies on *L. d. batavus* in the Netherlands, effort is now being concentrated on determining the feasibility of some larger fenland areas in the Norfolk Broads as sites for trial re-establishment. Conservation of the traditional habitats in Europe is a clear priority, in view of threats such as drainage, intensive fertiliser use, and mowing activities over the flight period (Ebert and Rennwald, 1991, quoted by Pullin *et al.*, 1995). The establishment of several populations in the UK sufficiently close to permit migration between them, is needed.

As Pullin *et al.* (1995) commented, the Large Copper conservation plan depends on close international cooperation, and it is sobering that conservation of this species in the UK — a programme which spans the whole recent scientific development of butterfly conservation — is still far from assured. The considerable effort devoted to this butterfly was summarized by Feltwell (1995).

Lange's Metalmark, *Apodemia mormo langei*

Lange's Metalmark was one of the first insects listed under the US *Endangered Species Act*, as one of several species found on the remnant riverine Antioch Dunes, California. Although the species itself is widely distributed in the western Nearctic, its colonies are usually isolated; the subspecies *langei* is represented by only a single population. The colony declined to a few hundred individuals in the early 1980s from the effects of habitat destruction through commercial sand mining; but over the last decade numbers have increased substantially in response to practical conservation management, to the extent that the butterfly may now be out of danger (Powell and Parker, 1993). It is thus an excellent example of successful conservation based on habitat rehabilitation, with good ecological data purposefully accrued for this (Arnold and Powell, 1983).

The butterfly remained abundant, despite the substantial sand mining reducing its habitat to two remnant patches by 1956. After 1972, increased mining on the western habitat and overgrazing by horses severely affected the colonies: mark-release-recapture studies from 1970 to 1982 showed a population decline from more than 2000 to fewer than 600 over that period.

The butterfly is univoltine, and populations are associated closely with the larval food plants (*Eriogonum*, Polygonaceae). Eggs are laid in small clusters on withering foliage low on the plant, and caterpillars hatch during winter; larvae feed until about July and then pupate in litter around the plant base. The plants senesce after 10–15 years, and the removal of plants by mining activities, and the prevention of recruitment into the *Eriogonum* population (in part through competition with abundant annual weeds, as the food plants reproduce only in open sandy environments), led to a substantial decline. Habitat restoration has involved extensive planting with *Eriogonum* (Powell and Parker, 1993) since 1979, as well as progressively securing the habitat against further human intrusions. In 1991, the creation of new sand dunes in previously mined areas commenced; these were later seeded and planted with *Eriogonum* and the important nectar plant *Senecio douglasii*, and subsequent plantings have helped to assure future supplies of larval food.

Powell and Parker (1993) commented that 'conservation and recovery efforts have been a dramatic success', but counselled the need for continuing regular plantings of *Eriogonum* to replace senescent patches as these become unsuitable. Because it is impracticable to prevent a ground cover of weeds which impede natural seedling growth, management involves clearing sites and planting new patches of food plant, rather than maintaining existing ones. Sustaining the butterfly will therefore depend largely on continuing human-aided replenishment of food plants.

The Palos Verde Blue, *Glaucopsyche lygdamus palosverdensis*

This subspecies was endemic on the Palos Verde Peninsula in California, and was recognized as endangered by the USFWS in 1980 because its coastal range scrub habitat was being destroyed or changed markedly by urban expansion (Arnold, 1986). Although three sites were designated as 'critical habitat' in 1980, loss continued, together with decline of the sole larval food plant (the Ocean Milk Vetch, *Astragalus trichopodus* var. *lonchus*) which is itself very patchily distributed. Arnold foreshadowed that the blue could become extinct very rapidly — and indeed, could have become so at the time he was writing — and thus be the first of nearly 350 United States organisms protected under the *Endangered Species Act* to do so. A recovery plan was approved in 1984 but its implementation was delayed considerably. Between 1984 and 1986 the number of *Astragalus* plants on the Peninsula declined from 318 to 80, and no life stages of the butterfly were observed for three years. By 1993, Mattoni (1993) commented that the subspecies was 'now certainly extinct'. However, its conservation value remained high as it had not been delisted, and development plans in the area must recognize habitat value for the butterfly, with consequent likely protection to other endangered species in the area.

Both the blue and its food plant naturally occurred at low densities and in very localized populations, so that both are highly vulnerable to various kinds of disturbance. The problems for management included small site size, degraded habitats, multiple ownership of the sites concerned, plans for future development and lack of funds for management activities. Arnold therefore recommended concentrating efforts at only five of the 10 extant sites, because dilution of available resources across all 10 would probably result in very little likelihood of success. Setting priorities for action and maximizing limited resources within a suite of priorities for action is a dilemma often faced in butterfly conservation, and the criteria are often difficult to assess completely in practical terms.

Arnold's priority sites were selected on (a) size (6–100 hectares), (b) 'naturalness' — all supported some native coastal range scrub vegetation, (c) ownership — four were owned by local government agencies and (d) zonation — four were zoned as open spaces by local municipalities, but part of the fifth was targeted for residential development. The other five sites would have been less suitable or politically accessible than these.

The Mission Blue, *Icaricia icarioides missionensis*

The Mission Blue is best known as the 'flagship' species for San Bruno Mountain, California, where two other butterflies (the San Francisco Silverspot, *Speyeria callippe callippe*, and the San Bruno Elfin, *Callophrys mossi bayensis*) and a number of other rare and endangered animals and plants also have their stronghold. It received federal protection as a declared endangered species in 1976 because much of its restricted habitat was scheduled for urban development, including the largest habitat remnant on San Bruno Mountain. This promoted one of the largest efforts yet made to conserve a butterfly. The costs of a large-scale biological study rendered this 'the most massive short-term study ever conducted on a butterfly in terms of both money and manpower' (Orsak, 1982). Thousands of butterflies were marked in 1980 and 1981 by a field crew totalling 76 people. Another dubious record held by the Mission Blue is as the endangered species which has halted the most expensive development — estimated in 1982 as US$8 million.

As well as on San Bruno Mountain, the Mission Blue occurs on Twin Peaks, about five miles to the north. Twin Peaks is the type locality of the subspecies, but housing development there reduced that colony to small remnants susceptible to trampling by sightseers, and San Bruno was seen as crucial for the butterfly's survival. It supported around 97% of the total population.

The major caterpillar food plants are three species of lupins (*Lupinus* spp.) which are also patchily distributed and generally rather sparse, and the habitat was also vulnerable to encroaching weeds and alien wildlife impact as a consequence of increasing nearby human populations. Trail bikes and other off-road vehicles were also involved, and changes in the overall species composition of the grassland were occurring. The lupin, although reasonably resistant to invasion by exotic grasses, is susceptible to shading caused by invasive woody plants (particularly gorse, broom and eucalypts on San Bruno Mountain). Controlling such plants on grassland sites is therefore important for the butterfly (Cushman, 1993).

The colonies of the Mission Blue tended to move from year to year, because only lupins of particular ages are utilized. Senescence of the lupins can lead to a decline in butterfly populations. More detailed knowledge of the ecology of the food plants was seen as a major priority need for planning effective management.

The total area of occupied habitat available on San Bruno was about 800 hectares — the San Bruno Elfin occupied only some 300 hectares — and many people opposed any development, believing that the whole area of 1460 hectares should be conserved for these two species and other unusual biota present. However, development of at least 13% of the total Misson Blue habitat seemed likely.

The detailed biological study led to a large three-volume 'Habitat Conservation Plan' and, unusually, this provided for an income of funding to be used for habitat management. It was proposed to reduce the extent of development of the area, so that prime habitat would be safe, and that buyers of new houses would pay an annual fee (proposed in 1983 as US$20 per house) as would tenants of commercial property (US$10 per 100 000 square feet rented). Both categories would be subject to indexation increase. This fund would at first yield about $60 000 a year and would be used for habitat maintenance, mainly by control of gorse (an invasive weed) and alien animals.

However, the compromise would be for the butterfly to lose part of its habitat to development — a political 'first' as the first case where part of the habitat of a listed endangered species in the United States would be destroyed with consent and agreement. Many conservationists expressed strong concern over the principle, especially when the habitat itself was limited and the ecology of the species concerned was not fully understood.

The major aims of the conservation plan were to enhance the quality of the habitat for the Mission Blue so that the effects of partial development could be countered effectively. Without any management, even without development or associated habitat alienation, it was considered that the species could still become extinct. Essentially, though, the plan satisfied nobody completely — the developers still had to spend a lot of money studying butterflies, they and local municipalities were concerned about delays and increasing costs of construction, and conservationists were not happy about any development of San Bruno Mountain.

The Small Ant-Blue, *Acrodipsas myrmecophila*

The Small Ant-Blue has been reported from a few scattered localities in eastern Australia, and is mainly known from hill-topping adults. It exemplifies two more general problems for conservation: needing to take notice of topographic features (hill-tops) used by adults for assembly, but without knowledge of any adjacent breeding habitat; and dealing with a species whose larvae are entirely (or almost entirely) myrmecophagous and spend their whole developmental period within the nests of a specific ant, which may itself be rare and elusive.

In Victoria, the butterfly has been reported in recent years only from one site, Mount Piper (80 km north of Melbourne), and both the assemblage (as 'Butterfly Community No. 1') and the species are legally protected under the State's *Flora and Fauna Guarantee Act.* Surveys to evaluate the status of the butterfly and host ant commenced there in 1991, but during the first three seasons only about five individual butterflies were recorded hill-topping, and attempts to find the sole recorded host ant in the vicinity failed (Britton *et al.*, 1995). A legal opportunity to define the 'critical habitat' for the butterfly therefore remained unfulfilled.

Earlier studies on a now-extinct colony of *A. myrmecophila* at another site in Victoria had revealed the host ant to be a species of 'coconut ant', a strong-smelling Dolichoderine of the *Papyrius nitidus* group, and many of the specimens now in collections were reared from caterpillars or pupae found in that nest. The ant nests in the ground and in dead timber, and steps were taken to deter the removal of dead timber for firewood from roadsides around Mount Piper and, by changes in land-use regulations, to reduce clearing of native vegetation on nearby properties flanking the central 56 hectares of Mount Piper, which was an Educational Reserve. Detailed searches for the ant involved pitfall trapping and direct searches, which collectively yielded about 150 ant morphospecies and confirmed the importance of the Mount Piper area as a reservoir of invertebrate biodiversity in a highly altered pastoral landscape (New *et al.*, 1996), but the conservation programme clearly depended on elucidating the biology of the ant and its interaction with the Small Ant-Blue.

In late 1994, Britton (1997) found a colony of *P. nitidus* on a private block abutting the Mount Piper reserve; it may have been established relatively recently, as earlier searches (and a pitfall trap grid operated only some 25 metres from the colony for several months) failed to reveal it. The colony was small, and dead timber was sparse on the open grassland site. To attempt to redress this, several artificial nest chambers, constructed from weathered wooden fence-posts, were placed on the site. They were colonized rapidly by ants, and early stages of *A. myrmecophila* were found within a few weeks. Continued monitoring of these trap nests has been instrumental in clarifying the butterfly's biology (New and Britton, 1977) and has had a major impact on the conservation programme at Mount Piper.

The site on which *Papyrius* was found has now been purchased and added to the core reserve. The use of trap nests enabled the butterfly's development to be monitored over several generations, and they could be used as translocation units to spread the ant and the butterfly over a larger site area, or to other historical localities. Work is also under way on the role of the powerful coconut scent of the ant as an attractant to the butterfly, in the hope that scent-based baits may help to overcome the highly expensive and uncertain process of monitoring or determining the presence of such elusive and apparently very rare butterflies with specific oviposition needs. The case of the Small Ant-Blue, which is still continuing, may therefore have important ramifications for other butterfly conservation studies.

The Eltham Copper, *Paralucia pyrodiscus lucida*

This lycaenid takes its name from an outer north-eastern suburb of Melbourne, Australia, from where it was described as a distinct race of the Dull Copper only in 1951. It was taken commonly for about a decade after its initial discovery in 1938, but declined markedly from about 1950 onwards and was feared to be extinct near Melbourne in recent years. An apparently thriving colony was found in early 1987, on land threatened with imminent housing development, and attempts to conserve the butterfly over the ensuing two years mark the first substantial efforts to target a butterfly for saving in Australia. The campaign has been a significant promotor of awareness of butterflies and other invertebrates

Figure 7.3 Male (top) and female (bottom) of an Australian Copper, *Paralucia pyrodiscus*, the species to which the Eltham Copper belongs (Anderson and Spry, 1893–4).

in the suburban community. The Eltham Copper has thus become an important flagship species for butterfly conservation in Australia, and is now one of the best-documented rarer Australian Lycaenidae. Its life history has been detailed by Braby (1990), and its distribution was discussed by Braby *et al.* (1992).

An urgent briefing paper to the State Government following the butterfly's discovery led to negotiations with the developers, who agreed to a moratorium on development of the site until the feasibility of raising funds for its purchase as a butterfly reserve could be investigated. The total cost of this was projected at A$1 million. The discovery of the Eltham Copper colony (Braby, 1990) coincided with the introduction of the Victorian Government's *Flora and Fauna Guarantee Act* (p. 83), and the butterfly was seen as a test of the State's sincerity in including invertebrates within the scope of this pioneering legislation. In conjunction with a public appeal, funding was provided to search for colonies of the butterfly elsewhere and to investigate its biology with a view to formulating a preliminary management plan (Crosby, 1988; Vaughan, 1987, 1988).

Ten discrete colonies are known around Eltham, and other colonies referred to this subspecies occur at Kiata and Castlemaine, both far away from Eltham in rural Victoria. The taxonomic status of the 'country' populations is not wholly clear, but there is no doubt that the butterfly is rare and has a highly disjunct distribution, and that the Eltham populations are separated from these by large areas of unsuitable terrain. Each of the Eltham colonies is isolated and the populations are largely or entirely closed. There is little likelihood of interchange of adults between most of the Eltham colonies and none over the broader range of the butterfly. Eight Eltham colonies are small and seem unlikely to be viable in the long term. The largest colony, still occupying only a few hundred square metres, was on the subdivision land and contained no more than 300–500 larvae, estimated directly by nocturnal counts. Only six butterflies were seen at the smallest colony. The intermediate-sized colonies were estimated to contain around 100–150 individuals. They are confined to open forest areas or degraded sites such as roadsides.

The only known larval food plant of *P.p. lucida* is a dwarfed form of Sweet Bursaria (*Bursaria spinosa*, Pittosporaceae), and caterpillars occur only on stunted plants associated with nest chambers of the ant *Notoncus*, to which the caterpillars retreat during the day. Decline is attributed to the destruction of native vegetation and habitat alienation associated with urbanization. The very small colony noted above is in a suburban garden.

The local community rapidly adopted the Eltham Copper as a local symbol for conservation. It became a familiar sight on T-shirts, badges and posters as the funding appeal got under way — with the appeal-line (around Australia's bicentennial year) 'Buy the butterfly a birthday'. The State Government pledged A$250 000 and the Eltham Shire Council committed a further A$125 000. The public appeal raised another A$56 000. Advantage was also taken of the campaign to foster interest in butterfly conservation through sales of a small booklet which put the Eltham Copper issue in a broader context (New, 1987). The total of A$426 000 in 'real money' is by far the largest sum ever committed to the purchase of a habitat specifically for a butterfly in Australia, and seems unlikely to be surpassed. The value of this was augmented considerably by the State government transfer of an area of land, near the main subdivision colony and which supported another substantial colony, to constitute part of the butterfly reserve.

It is still too early to determine whether the Eltham Copper will be conserved satisfactorily; part of the subdivision site has been reserved, but the small size of the buffer around this, together with the intrusion of housing on the remainder, may result in further decline. Management recommendations for the Eltham Copper (Vaughan, 1987, 1988) included:

- protecting all suburban colonies from the various threatening processes associated with urbanization — such as trampling of vegetation, garbage dumping, sullage overflow, weed invasion and activities of exotic animals,
- providing for habitat expansion by promoting regeneration, or propagation and transplantation, of *Bursaria* from any sites clearly to be destroyed by development, and
- providing for a ranger to foster practical management and monitor its effects.

The butterfly has been monitored continuously since 1991, with the help of a group of concerned people, the 'Friends of the Eltham Copper', and in recent years counts of caterpillars and butterflies have been made (Braby *et al.*, 1996; van Praagh, 1996). Weed invasion is still a principal concern in habitat degradation, and management for the *Bursaria* in relatively open areas is a matter of considerable importance as the sites mature and trees grow to produce more shade. Increasing urban pressures are still evident, and the accumulation of vegetation fuel is of concern in this fire-prone area. It is anticipated that one key colony site will be subjected to a control burn in 1997 as a trial to determine the optimum methods for site maintenance. Vaughan's (1988) recommendations remain equally relevant nearly a decade later.

The Atala, *Eumaeus atala florida*

The Atala occurs in southern Florida, and was formerly abundant there. However, urbanization and coastal development, together with large-scale havesting of its cycad food plant (*Zamia pumila*), reduced it to a single known population by 1965 (Emmel and Minno, 1993). This colony died out, and the butterfly was feared extinct.

The Atala, which gave its name to a journal formerly published by the Xerces Society, is the largest lycaenid in Florida and is closely associated with its sole larval food plant. It received full species-level treatment in the IUCN Invertebrate Red Data Book (Wells *et al.*, 1983), in which it was categorized as 'vulnerable', following its expansion from a colony found near Miami in the late 1970s. It is not yet clear whether this Atala is indeed the same taxon present before 1965, or the closely similar nominate subspecies from the Bahamas. From this colony, the species expanded substantially and recolonized (or has been translocated to) much of its former range, in conjunction with widespread plantings of *Zamia*. It is thus now out of danger. Indeed, it is now considered a pest species by some commercial nurseries involved in rearing cycads for ornamental use and landscaping, and by botanic gardens.

However, there is little doubt that some mainland populations were harmed by a hurricane in August 1992 (Emmel and Minno, 1993), and the species is of concern because of its vulnerability to such sporadic events, as well as to pest control measures in urban environments and continued destruction of natural habitat.

CHAPTER EIGHT

BUTTERFLIES IN TOWNS AND GARDENS

Several of the cases discussed in the last chapter referred to butterflies in, or close to, urban environments. This chapter is a broader treatment of 'urban butterflies', and includes advice on the steps which individual people (in contrast to the more organized 'scientific' approach) can take to foster the wellbeing of butterflies in home gardens.

We tend to think of butterflies as 'country animals' and, indeed, I have stressed earlier the fact that urbanization has led to the decline of many species. Older collectors, in many parts of the world, have vivid and fond memories of favourite collecting localities in areas which are now covered with concrete or asphalt. But cities and towns contain open spaces with vegetation; parks and similar areas play important roles as oases for remnant wildlife, and there are many ways in which suburban gardens can be used as a key resource in the conservation of butterflies. In the UK, Owen (1978) claimed that gardens together occupied an area about 10 times as great as that of national nature reserves, and more species of butterflies were recorded from his midland garden than from any local nature reserve.

Generally, the species present in towns are a restricted subset of those which would occur naturally in the region. Many of the references to urban butterflies are little more than lists of species, sometimes of impressive length, but there have been a few more penetrating attempts to allocate such butterflies to ecological groups or to determine why they are able to persist.

Shapiro and Shapiro (1973) assessed the butterflies of Staten Island, New York, and found that the relatively few species on abandoned ground on the urbanized north of the island were increasing in distribution and abundance, whereas specialized native species were declining. The 'successful' species here included easily dispersing colonizer species with a high reproductive rate, and caterpillars which often fed on weeds. They could also, probably, tolerate a level of air pollution. In Sapporo, Japan, Yamamoto (1977) demonstrated a substitution of forest species by more open ground species. Few species were common, and these, tending to have at least three generations each year and to overwinter as pupae, were considered relatively 'resistant' to urbanization. Much earlier in the UK, Owen (1954) had drawn attention to then-abandoned bomb sites in London as valuable habitats for butterflies. And in Porto Alegre (Brazil), Ruszczyk (1986, 1987) showed that butterflies were well-correlated with the 'urbanization gradient'. Increase in urbanization and pollution was accompanied by decrease in numbers of both species and individuals. Species typical of open areas, again of high vagility and with larvae able to feed on exotic cultivated plants, were dominant in the city. Counts based along major routes radiating outwards from the city showed a clear transition to more 'typical' fauna, as distance reflected such important parameters as climate, percentage of vegetation cover, pollution and human density. The more detailed analysis by Ruszczyk and de Araujo (1992) confirmed that spatial patterns of butterfly diversity (Shannon–Weaver index) and species richness were strongly correlated with environmental variables, including a variety of abiotic factors.

Urban gradients in California also showed changes in butterfly species composition, reflecting the extent of change from the former oak woodland (Blair and Launer, 1997), and such patterns may be widespread as a reflection of habitat change. In the Californian case, any development was apparently detrimental to the original butterfly assemblage, so that the greatest effective conservation for butterflies would be derived from concentrating development into the smallest areas needed and keeping other land as close to natural as possible.

However, there has been considerable recent attention paid to the management of 'urban reserves' for the wellbeing of butterflies, with the value of small uncommitted areas such as roadsides, parks and river banks being increasingly recognized (Singer and Gilbert, 1978). Not only is the value of native vegetation, including weed species, now widely acknowledged, but the possible augmentation of butterflies by deliberately planting larval food plants and attractant nectar plants as adult food supplies is also becoming more common. Individual people can play vital roles in these important aspects of conservation through their home gardening practices and by influencing local councils in their treatment of nature strips, road divides, parklands and other open spaces. Competitions for 'tidy towns', 'best-kept village' and the like are now commonplace in many developed parts of the world but, whereas these formerly encouraged sanitation in the sense of promoting neatly mown verges and the removal of all weeds, pondside vegetation and so on, this attitude is now changing. Wildlife resources such as some weed patches and untrimmed hedges are now — at least in some places — valued in such competitions and gain appreciation from enlightened judges. Such changes in attitude are likely to lead to the conservation of many breeding habitats for butterflies.

These changes are probably not restricted to temperate region butterflies, although these are the best understood. Owen (1971b) commented that a considerable proportion of common African butterflies utilize agricultural weeds as larval food plants. As a result they have expanded their range as forests have given way to cultivation. In some cases they seem to be adopting previously unused food plants.

Tropical gardens in many parts of the world tend to resemble each other, because ornamental trees (often with conspicuous flowers or decorative foliage) have been exchanged between various widely separated places (Owen, 1971b). Species diversity of butterflies in a tropical garden can be very high. Owen's garden in Sierra Leone yielded more butterflies than secondary or primary forest because it attracted many typical savannah species which were absent from the forests. A number of forest species, including genera characteristic of the forest floor, were also recorded. According to Owen (1971b), tropical gardens tend to have: few of the original trees typical of the region; ornamental trees, including exotic species from other parts of the world; a lawn or grassy area; a variety of flowers providing blossom throughout the year; and weeds. However, from the conservationist's point of view, gardens may harbour many 'generalist' butterfly species but few of the ecological specialists of greatest conservation interest.

BUTTERFLY 'GARDENING'

With care and knowledge, deliberate seeding or planting of gardens to attract butterflies can be undertaken, and a number of plants have excellent and well-founded reputations for their ability to do this. *Buddleia*, for example, is commonly referred to as 'the butterfly bush'. Shaded areas are important, as well as some patches of native grasses. The principle of having a small area of 'wild' garden — where weeds and/or native vegetation can flourish — is also well-established and important in fostering the conservation ethic. Newman (1967) wrote at the start of his book *Create a Butterfly Garden* 'All man-made gardens are artificial to some extent and the more we try to impose order and tidiness on our surroundings the more we are interfering with nature'. He used the example, for the UK, that a number of the more conspicuous nymphalines have caterpillars which feed on stinging nettles, weeds associated with human habitation but generally rapidly subjected to weedkillers and other forms of elimination. A combination of a small nettle bed and nectar flowers is sufficient to maintain garden populations of several species of these highly coloured butterflies. A weed, of course, is only called that because it grows where it is normally not wanted, and rapidly loses any connotation of undesirability if it becomes welcome. Small patches of milkweeds (*Asclepias*) in gardens over much of central eastern Australia may well encourage oviposition by Wanderers, which would otherwise only be passing through the area. Both domestic gardens and larger areas such as managed urban or suburban parks and botanic gardens may provide valuable habitat for many butterflies, albeit on rather different scales. Management of such areas to encourage butterflies has become more frequent, and in the UK a national garden butterfly survey was launched in 1990 with the objective of discovering how butterflies may be fostered in gardens and as the start of a survey of year-to-year changes in butterflies in gardens (Vickery, 1995). Vickery also summarized a number of earlier UK surveys for garden butterflies, but more than 1000 gardens were included in the recent survey. Recorders are asked to complete two forms. One deals with butterflies, listing 19 species recorded as visiting gardens, but with provision for adding others; the second surveys the common nectar plants present. Features of the garden, such as size and aspect, are also reported.

In general, larger gardens contained more butterfly species than medium or small gardens, and diversity also increased with proximity to natural habitats, so that rural gardens attracted more species than urban or suburban ones. Most records were of adults, and relatively few records of breeding were accrued. Vickery (1995) emphasized the great value of gardens as sources of nectar for butterflies, especially in spring and autumn when wild plants used as nectar sources may be scarce. In dry seasons, also, nectar supplies in gardens are fostered by watering (Vickery, 1996).

Comments on the behaviour of butterflies in a South African botanic garden (Wood and Samways, 1991) reflect how landscape elements can influence abundance and diversity. As is the case for many other urban oases and reserves throughout the world, there are very few lists of which species are present and resident in most botanic gardens. Enhancing the nectar supply and planning adequately for a range of environments, such as different degrees of shade, can be important in fostering a range of butterfly species; careful consideration may be needed if introducing exotic plant weeds which might have adverse effects on other native flora and fauna (Feltham, 1995). 'Butterfly gardening' has been defined as 'the art of growing plants that will attract butterflies' (Booth and Allen, 1990); though this is clearly a central need, the somewhat different parameters of microclimate and topographic needs cannot be neglected if resident butterfly populations are to be sustained in relatively small, isolated, and highly managed areas.

Thistles, blackberries, ragwort and other 'weeds' are recognized as flowers which attract adult butterflies to feed on them, and children can sometimes be awakened to a life-long interest in natural history by having an area of a garden in which they can sow seeds of butterfly plants collected from the wild. Packets of 'butterfly plant seeds' are sold in the UK, and could become marketable elsewhere. Some are now available in Australia. Many flowers attractive to gardeners and horticulturists are *not* attractive to butterflies! Many herbs and open-flowered composites (daisies, and the like) are far better for butterflies than the largest or grandest roses, lilies or chrysanthemums. With care, an array of flowers can be prepared which will attract butterflies throughout the spring and summer in temperate regions and the whole year in the tropics. The basic needs are nectar, scent and bright colours, and observations will quickly show that particular colours may predominantly attract particular kinds of butterflies. The general themes of 'butterfly gardening' are explored in several recent books — that by Rothschild and Farrell (1983) is very informative, and a recent compendium by the Xerces Society (1990) also contains much practical advice, particularly for North America.

Additionally, it is usually possible to provide food for butterflies — a sort of 'insect bird-table' — which they visit in order to feed. A weak mixture of honey in water (one widely disseminated recipe cites half a teaspoon of honey with half a teaspoon of castor sugar and a small pinch of salt, all in a cup of water) can be provided in a shallow dish or on pads of cotton wool. If the solution is too strong the butterflies may stick to it, so that over-weak solutions are far better than more concentrated ones. Adult butterflies detect such sweet liquids by sensory structures on their tarsi (the end part of their legs), so they sometimes need to tread on the food once they have been attracted to it by scent. One problem with exposing such foods is that the butterflies may then attract birds which feed on them. To counter this, Whalley (1980) suggested suspending cotton wool pads from branches in the garden. Newman (1967) advocated suspending test tubes of feeding solution plugged with cotton wool (Figure 8.1, p. 191) from plants. The attractiveness of these can be enhanced by a red, purple or blue paper 'collar' on which the butterflies can rest while they feed.

This 'artificial flower' approach will be very familiar to people accustomed to providing food for nectarivorous birds, and can be used in various ways. Two devices formerly used in the Melbourne Zoo's butterfly house, for example, consist of (a) coloured hollow plastic golf tees pierced through the centre of dried 'everlasting' flowers and held in blocks of expanded polystyrene foam and filled with food and (b) coloured plastic beads of the children's pop-necklace kind in saucers of food filled to about half the diameter of the beads (Figure 8.1). Other equally ingenious and simple options will surely suggest themselves, and the great variety of artificial flowers now easily available may be employed as feeders in this way. 'The butterfly gardener must gather his or her own information through careful experimentation and observation' (Jackson, 1986).

REARING AND MAINTAINING BUTTERFLIES

Many people also like to rear butterflies in captivity for release into their gardens, and find this an instructive and fascinating hobby, commonly viewed

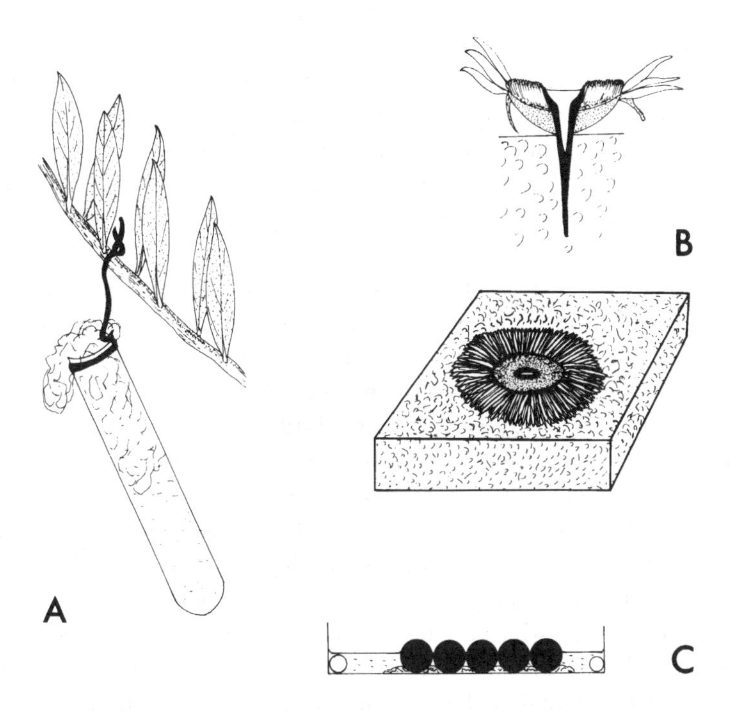

Figure 8.1 Some feeding devices used for adult butterflies: A, a tube of honey-water solution plugged with cotton wool and suspended from vegetation; B, everlasting flowers pierced with a golf tee (see section) into expanded polystyrene; the well of the tee contains food solution; C, coloured plastic beads supported as a feeding platform in a petri dish of food solution.

as an aesthetic exercise rather than as 'conservation'. But doing this carefully, keeping notes or a diary associated with it, can contribute to the broad information base needed to understand butterfly natural history. Captured female butterflies will often lay eggs readily in captivity. They can be housed in relatively simple cages with honey-water solution as food and suitable plant material on which to oviposit. Some butterflies will lay on the sides of a cage, but others need the precise larval food plant, so it is helpful to provide this wherever possible. It is sometimes worthwhile to induce captive butterflies to feed through 'force-feeding', especially if they have been confined for several hours after being captured. Carefully-netted butterflies can be transported easily, with their wings folded, in folded triangular paper envelopes which prevent them moving around and becoming agitated. A small twist of paper tissue or cotton wool will prevent them from being crushed inadvertently. (Similar packaging can be used for sending live butterflies by post; the envelopes can be packed loosely in rigid cardboard boxes for this.) For bringing them back from the field they can also be placed singly in small plastic or glass-capped vials with paper for them to rest on, but these are considerably more bulky than envelopes if many specimens are involved. In hot weather, containers of live butterflies should be transported in a polythene cooler, Esky or similar container, to prevent over-heating.

Force-feeding (Figure 8.2) involves holding the butterfly carefully by the wings or body between finger and thumb and gently drawing out its coiled proboscis with a fine pin onto the surface of a feeding pad by drawing the pin gradually away from the insect. Initial contact with food will often induce the butterfly to start feeding, but several attempts may be needed for some individuals.

Cages for obtaining eggs from mated butterflies need not be large. Indeed, some of them will lay in vials or boxes that are too small for them to stretch their wings fully. Because of the courtship aerobatics practised by many species, much larger cages may be necessary — the main hindrance to establishing a captive breeding operation for Queen Alexandra's Birdwing, for example, is the initial cost of very large flight cages believed to be needed for mating. Some other swallowtails can be hand-paired in captivity, and for many other butterflies this space problem can be overcome (at least for the amateur collector or enthusiast working with more common species) by selecting 'fat' looking females which are likely to have mated already.

Examples of cages which can be used for butterflies and for subsequent rearing of caterpillars are shown in Figure 8.3. These are all widely used patterns but their presentation here should not preclude ideas for others. The 'box' cage with mesh sides and wood, chipboard or laminate floor can be made rot-proof with selection of stronger materials for outdoor use, and allows free passage of air to prevent condensation. Dimensions can be varied according to space available. The 'cylinder' cage can be made from discarded biscuit tins or baking trays, as the base and top, or rust-proof equivalents can be purchased from dealers in

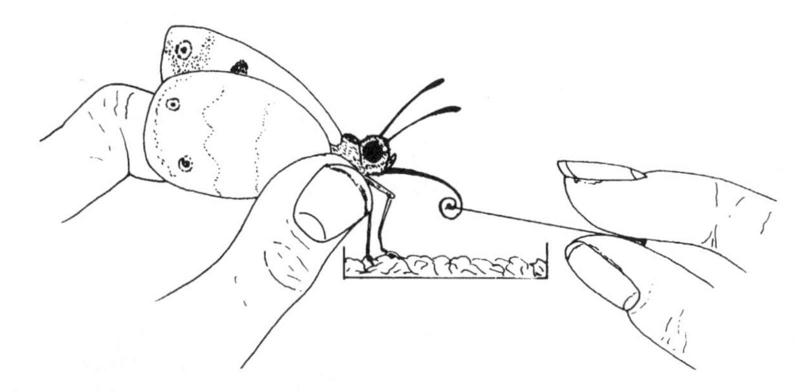

Figure 8.2 Hand-feeding a butterfly. The butterfly is held gently between finger and thumb (or with soft forceps) and allowed to rest on a cotton pad of food solution. The coiled proboscis is gently extended with a fine pin and moved onto the food. Several attempts may be needed before the insect starts to feed.

entomological equipment. The lid and base are separated by a cylinder of a clear rigid plastic (cellulose acetate), which can have muslin-covered air vents, and the lid should be provided with similar holes to aid ventilation. If used outside, drainage holes in the base should also be provided. The third pattern is ideal for

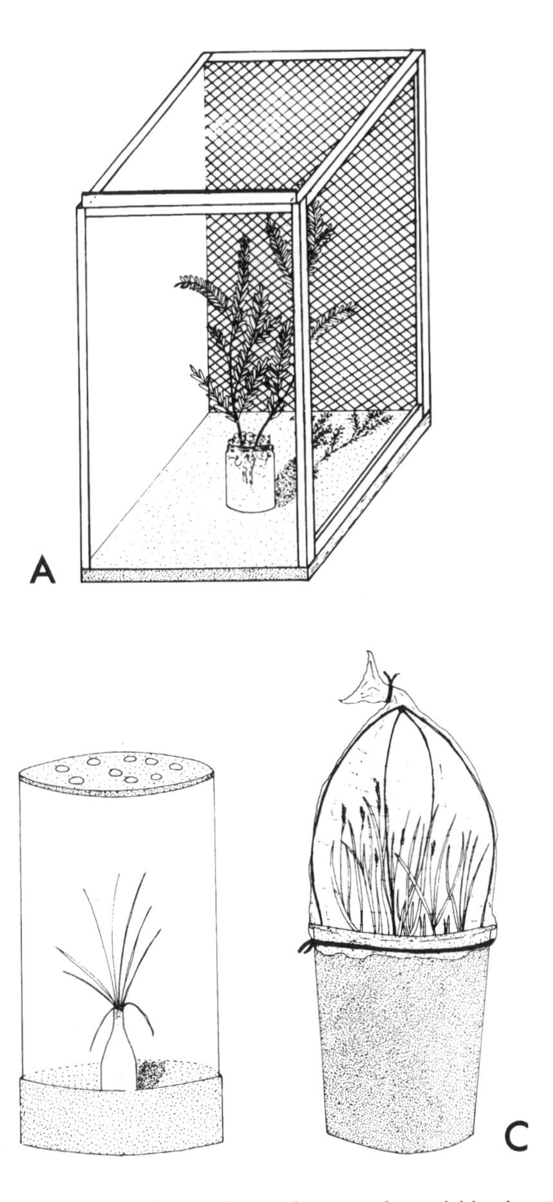

Figure 8.3 Three patterns of rearing cages for caterpillars: A, a box cage, with vertical sliding door; B, a plastic cylinder supported by metal base and roof; C, a plant pot covered with a leg of pantihose supported on a wire frame, which can be made from wire coat hangers. A and B are used for cut plants, which are supported in containers of water, and C is useful for growing plants.

small potted plants and consists simply of a stocking or the leg of a pair of tights stretched over a wire frame which can be made easily from coat hangers. This pattern can be extended onto large plants by 'sleeving', using large weather-resistant mesh bags or cylinders tied over branches of trees (Figure 8.4), or even whole small bushes. Sleeves may be vulnerable to invasion by predators and parasites but have the advantage of relatively low maintenance for rearing larger numbers of caterpillars. As with any rearing enclosure, though, cleanliness is very important, and frequent checks are needed to detect early incidence of disease or depletion of food plants.

Many caterpillars are best kept in small groups while young, in plastic sandwich boxes or similar clear plastic relatively airtight containers, with food provided as single leaves or in small amounts every day or so, placed on tissue or absorbent paper. At this stage, especially, caterpillars are very delicate and are best 'handled' using a small soft paint brush rather than directly by forceps or fingers. Details of the food plants needed for many butterflies can usually be obtained from local handbooks, or good 'leads' obtained by seeing what the

Figure 8.4 A sleeve cage. For rearing caterpillars outside, a muslin bag or 'sleeve' is tied over a branch or plant to enclose them.

species' close relatives eat. With some exceptions, Lycaenidae are particularly difficult to rear in captivity because of their additional complex relationships with ants (p. 25).

There can, at times, be problems with supplying the correct food, either because it is not known what particular caterpillars eat, because particular food plants are not available, or because the identity of field-collected caterpillars is not known. Especially if large numbers of caterpillars are to be reared, the use of artificial diets may be warranted. Morton (1981) reviewed the early use of these for caterpillars, pointing out that they have historically been much more widespread for rearing moths than butterflies, and gave a recipe for a diet on which he reared about 50 species of butterflies, including representatives of most families. Details of diets are found also in a number of books on rearing Lepidoptera, such as Dickson (1976). It is usually necessary to include a proportion of dried larval food plant because chemicals peculiar to it may be needed to stimulate caterpillars to feed. For larvae whose precise natural food plant is unknown, it may be possible to 'test' diets containing foods of related species — many Satyrinae, for example, will feed on a number of different grasses. As well as its composition, the texture of the diet may be important. It must be solid, so that caterpillars do not drown in liquids, but not too hard for them to chew. Details for making Morton's diet are given in Appendix 3. Such diets can be stored for quite a long time at around 4°C, and can be provided to caterpillars in jars or disposable plastic drinking cups. Their acceptance by newly hatched caterpillars, which commonly wander around for a day or so before they start to feed, varies from about 30 to 100%. Morton (1981) also found Lycaenidae the most difficult family to rear.

When fully grown, caterpillars of many butterflies wander away from their food plants to pupate. They may move downwards into ground litter or seek places from which to suspend themselves. It may well be necessary to deliberately supply these at the time needed. Where possible, pupae should not be handled until they have become firm, and not at all if the cage can be spared to leave them undisturbed. If pupae are on the cage floor, the provision of a rough surface (such as corrugated or rough card) will enable emerging adults to grip in order to expand their wings, and it may be necessary to counter desiccation by an occasional light spray of water. Without this, some 'crippled' adults which cannot expand their wings fully are likely to result. If suspended pupae must be removed or are brought in from field collections, successful emergence can be enhanced by again hanging them up rather than just leaving them on the cage floor.

Adult butterflies should be released only when their wings have hardened fully, because before this they are very vulnerable to birds or other predators. They should be released in warm, calm weather, preferably close to the plants on which it is hoped they will feed and away from the attentions of birds or cats.

They may benefit from a pre-release feed but, of course, there is little control over most butterflies once they have been released and they may then disperse well away from the point of release. Some people prefer to release butterflies into a muslin or mesh 'tent' over flowers initially, in the hope that they will feed naturally and become habituated to their surroundings. These may be particularly important in conservation releases, to facilitate reproduction before dispersal reduces the chances of small numbers meeting to mate.

It is important that such releases are made only within the natural range of the butterfly. Apart from ecological considerations (alpine butterflies are not likely to thrive in lowland gardens, nor deep forest species in open areas) and the controversy over 'reintroductions' we noted earlier (p. 132), the most successful introductions into the suburban gardens are likely to be those very species which thrive in the area and may already be present, albeit in small numbers. The primary aim should not be to introduce exotic species, but to increase the population levels of the species typical of the area.

Reduction or elimination of pesticide use is also an important consideration, not only for butterflies but for other beneficial insects such as bees and various other pollinators and predators. Most common garden insecticides are broad spectrum and will kill many kinds of insects and related animals indiscriminately. Although there has been a widespread swing away from pesticide usage by many gardeners, the levels applied are often still high — often higher than those used by commercial growers on an area-for-area basis. 'Natural' pest control is now more popular than ever before, and there are a number of worthwhile and comprehensive accounts of this (for example, Chapman *et al.*, 1986). Few garden pests are butterflies, but measures taken against other pests may harm butterfly caterpillars, and weedicides can destroy their food plants. Where needed, these should be applied only in calm conditions, to avoid any drift. There are often viable alternatives, however. Soapy water can be used instead of insecticides, for example, to wash many pests (such as aphids and thrips) off plants. For vegetables, some protection from pest attack can often be engendered by planting different kinds in single rows or small patches, rather than larger stands which present invading insects with an easily identifiable target. One reason that crop monocultures are vulnerable to pest attack is that they are large areas of low diversity vegetation which consumers can readily colonize and exploit.

Some form of companion planting has high credibility as a mode of protection against pests, but the reasons for this are generally not fully understood. Essentially, the addition of other kinds of plants increases the diversity of vegetation, and even weeds can be beneficial! Several studies on *Brassica* crops, for instance, have shown that fewer white butterfly caterpillars occurred on cabbages grown on weedy plots. Many of the more elaborate repellent quality claims made for companion plants are very difficult to substantiate but, again, the value of planting nasturtiums, marigolds or catnip among cabbages to reduce *Pieris* attack seems to be well-established, at least on some occasions. However, the cabbages still suffer some loss of yield compared with fully 'controlled' pest situations, so this strategy may not be acceptable for commercial growers. Latheef and Ortiz (1983) summarized much of the early information on companion planting and *P. rapae* attack in North America. Results for some plants are decidedly contradictory and, in some cases, interplanting with herb mixtures resulted in *increased* attack over levels in monocultures. Surprisingly, marigold, nasturtium, peppermint, sage and thyme interplanted with cabbage seemed to attract *P. rapae*.

There are occasional cases of butterflies being introduced into gardens or parks inadvertently, though such events are difficult to document clearly in retrospect, or to separate from unusual but natural dispersal. Some of the turf-feeding skippers may be particularly prone to this. They include, as possibilities, *Thymelicus lineola* (the Essex Skipper) in the UK, and now a pest of some grasslands in Canada, and *Ocybadistes sothis walkeri* (the Yellow-banded Dart) in southern Australia. The latter has spread rapidly around Melbourne during the last few years, and Crosby and Dunn (1989) suggested that it might have been introduced to the area in lawn turf rolls grown elsewhere. Some of its preferred grass foods have been used extensively for lawns in public areas such as schools and sports grounds during recent years, and much of this was imported from within the former natural range of the skipper. It is very easy when bringing plants into a garden to also bring in eggs or caterpillars associated with them, especially if they have not been grown from seed in nursery conditions.

Brewer (1982) suggested several simple 'rules' for butterfly gardening, and these summarize a great deal of common sense:

- Know what you already have. A butterfly garden may necessitate at least three years of observation, trial and error planning and planting. With knowledge of species which occur naturally in the neighbourhood and of their larval food plants and preferred nectar plants, it should be possible to plan constructively rather than haphazardly.
- Design around the sun. While watching butterflies, notice the position of the sun in relation to the flowers on which they rest. Often, the difference between flowers visited and those of the same species which are not is a difference in orientation or exposure to the sun. Planting can be planned accordingly, and inadvertent shading (such as from planting under trees) can be minimized.
- Keep it simple. Concentrate on augmenting and encouraging species which occur naturally nearby. Planting larval food plants may encourage oviposition and if more females can be attracted by providing good nectar sources, more eggs (and caterpillars) may ensue. Normally, nectar flowers will not

bring in butterflies from more than the distance they would usually fly in seeking food. While exotic flowers (such as *Buddleia*) will normally yield good results, the preferred flowers are often those which are native to the area and with which the butterflies are (as species) familiar.

- Adding plants or new butterfly species to the garden should be secondary to encouraging local species which can thrive in the area on foods which are available locally.

Measures suggested to encourage butterflies on farms (FWAG, 1995) are applicable in principle to gardens and other similar areas. These stress the dual approach of maintaining or improving habitats and reducing operations likely to harm them. The former encompasses need for sun, shelter, nectar and food plants. The latter involves avoiding unnecessary chemical spraying, with total avoidance of spray in areas such as borders, hedge bottoms and buffer areas, avoiding spray drift onto such regions, and accepting some 'untidiness'.

CHAPTER NINE

THE FUTURE

There is no general 'formula' for saving threatened species of butterflies. The series of cases outlined in Chapter 7, together with others, do have elements in common, and these can be synthesized to some extent to produce a scheme of practicalities which is, at the least, a sequence of useful 'rules' or 'points to consider'. This is based on a pro-forma designed by Arnold (1983) and relies on the following sequence:

- ensuring the habitat is safe from despoliation;
- progressively documenting the butterfly and its ecological requirements so as to move towards habitat maintenance and improvement;
- monitoring the population to determine the success of this, and continually refining management in the light of new information;
- informing and educating people and evaluating any form of legal protection.

In principle this sounds straightforward; in practice it can be immensely complicated and difficult to achieve, and so is worth looking at in more detail.

Habitat preservation and protection. The initial step of preservation is often difficult, not least because preservation often entails a conflict over the use of the site concerned — often, for 'crisis management' conservation attempts, with commercial interests. It may necessitate site acquisition by donation or purchase (perhaps at an inflated value), changes in status for public land to enable site reservation, establishment of conservation easements, or co-operative agreements with land owners or managers. Subsequent protection of larval and adult resources may involve substantial control of access to the site. This may include such factors as:

- minimizing the uses of pesticides and/or other toxins;
- minimizing uncontrolled intrusion by domestic stock, humans and off-road vehicles to reduce or eliminate trampling or destruction of food plants or other vegetation;
- preventing intrusions of exotic plants; and
- preventing or minimizing direct removal of native vegetation, ranging from herbs to trees, so that land clearing is forbidden or strictly controlled.

The first of these may cause difficulty over what constitutes a 'pesticide'. Diseases, such as the commercial preparations of *Bacillus thuringiensis* commonly used against pest Lepidoptera, are commonly regarded as 'safe' because they are chemically non-polluting. But drift of *B. thuringiensis* onto an area used to conserve butterflies could cause major havoc; Kudrna (1986) advocated a complete ban on its use in areas inhabited by significant butterflies, and a broad 'safety belt' around sites. He also cautioned on the use of fungicides — although their effects are sometimes not well understood, they may alter host plant quality and acceptability to caterpillars. In general, buffer areas to help counter disturbance are highly desirable. Effective status of habitats can sometimes be increased to 'critical habitats', as in the United States.

In cases where the butterfly concerned is of ambiguous or unclear taxonomic status, it may be necessary to clarify this in relation to other or outlier populations in order to determine whether it is as unusual as first suspected. In general, habitat reservation is 'easier' (in that more persuasive cases can often be made) for taxonomically unique populations than for isolated populations of more widespread taxa. Even if the butterfly is rare or unusual, the very fact that it does occur elsewhere may be taken to diminish the conservation significance of any one site. However, regional or national efforts often do not coincide with target taxa which would be identified by a broader consensus. As we noted earlier, the massive efforts in the UK have been directed mainly at species which are widespread elsewhere in Europe, but on the edges of their ranges in the UK (p. 126).

Habitat maintenance and improvement. The aim is to maintain and improve the habitat in relation to the requirements of the butterfly. Various habitat improvement methods may be obvious and relatively simple on the basis of knowledge already available on the butterfly. Examples of this could be removal or control of exotic weeds or noxious plants to render food plants more vigorous or accessible, promotion of food plants and other natural vegetation — a gradual process of restoration which may entail propagation and transplantation in conjunction with weed removal, and promotion of particular management tactics such as grazing regimes for grasslands, flood levels for marshes, and coppicing for some woodlands. It is important to simultaneously attempt to determine (or refine knowledge of) the optimum physical and climatic factors beneficial to the species and relate this to local habitat enhancement to provide 'foci' within the overall site. Knowledge of the ecology of the target species is the key to rational and efficient management.

Many aspects of the butterfly's biology may need to be studied more intensively, depending on how much information is initially available. They include the life history and pattern of seasonal development, dependence on particular plant species or growth stages or (Lycaenidae) on particular ants or Homoptera, the status and degree of isolation of the populations, adult behaviour (the factors influencing successful mating and oviposition), predators and parasites and other factors which either cause direct mortality or counter natural population growth, all of which can add to the effectiveness of management and help to predict the consequences of particular endangering processes or events. It may also be worth exploring the feasibility and potential value of captive rearing for population augmentation, or even range extension.

Additional knowledge of food plant ecology can also be valuable, as a basis for providing optimum plants for caterpillars or sources of nectar for adults. In the case of caterpillar food plants, knowledge of reproductive processes, what other animals exploit or hinder its growth, and whether particular soil, exposure, slope and other factors foster its wellbeing, can be invaluable. Horticultural studies to assess propagation techniques for augmenting field stocks can also prove useful in particular instances. All these factors may need to be evaluated for incorporation into development of a long-term management plan.

A further problem arises when different butterfly species in the same habitat have different, contrasting, requirements. In Chapter 7 we discussed the Adonis Blue on chalk grassland in the UK. However, different butterfly species there prefer different sward heights, and a survey of their requirements (BUTT, 1986) led to some general principles of butterfly management, and emphasized that a mosaic of microhabitats is usually needed within the gross structure of a site to foster continued co-existence and wellbeing of species with disparate ecologies. For grassland, grazing regimes can be critical for maintaining particular sward characteristics, and understanding the effects on vegetation structure of continuous versus sporadic or rotational grazing, and of the kind of animal used, is important. Mowing and burning may, in some cases, be projected as management alternatives to grazing, but neither maintains an equivalent mosaic of microhabitats or vegetational complexity.

Monitoring and management. Monitoring is needed to determine the success of management and to modify this as needed. It may be necessary to modify or develop methods to estimate population numbers and trends in the abundance of butterflies and their caterpillars, and to select particular sites or subsites for relatively intensive investigation.

Education. Throughout all of this it may also be necessary to increase awareness of the species among the public and decision-makers by developing education or information programmes. These may include information signs (unless there is a perceived risk of these increasing levels of unwanted collecting or intrusion), possibly interpretative tours for naturalist groups, school children and visitors (if the site will permit this without causing or exacerbating problems), media interviews, publications, newspaper articles on the significance of the butterfly and how it is being managed, and continued liaison with all interested parties. If any additional legal protection is needed, this can also be promoted, and any available regulations can be enforced.

It is vital to maintain and foster public goodwill in conservation programmes of any sort — but in sensitive cases concerning rare butterflies the balance between the positive values of publicity and the possible dangers of overexposure is a very fine one. A very few extra people trampling a site or generally responsible collectors taking 'just one specimen' for their personal collections could make all the difference to a highly vulnerable population.

The essence of this sort of scheme is to provide some kind of rational framework for butterfly conservation attempts, which are often 'last minute attempts . . . carried out haphazardly in the race against time' (Kudrna, 1986). It is obviously utopian, and I can envisage many experienced practitioners of butterfly conservation shaking their heads sadly at the sheer impracticality of prosecuting a full scheme of this sort, because of time, lack of funds, and logistic constraints.

There is, though, an increasing appreciation of the need for comprehensive conservation programmes for butterflies, and of the need to seek informative generalities across a wide diversity of cases. In addition to emphasizing the needs for broad surveys of butterfly distributions in Europe, Kudrna (1986) has emphasized that different ecological categories of butterflies will need rather different approaches. He proposed a conservation programme for European butterflies, and distinguished between the needs of:

- species acutely threatened by extinction;
- rare endemic species;
- eurychoric species; and
- incipient species.

Taking these in turn, the major thrust of conservation for each was considered.

Acutely threatened species. As all species acutely threatened in Europe are ecological specialists, the major initial step must be to halt further decline by safeguarding the existing colonies. Progression towards documentation and management would entail mapping of all inhabited sites and adequate buffer zones, and assessment of their susceptibility/sensitivity to anthropogenic influences; recording of past and present management; exhaustive ecological study of the species to provide information for specific management; monitoring in selected sites and, if necessary, full legal protection of some sites, with restrictions on collecting if these should become necessary.

Rare endemic species. Many rare endemic species (for example, more than a third of Europe's indigenous butterflies) have very restricted habitats. Two major categories were recognized: those which are known from one or few localities, and those which are more widespread. The former are particularly susceptible to human influences, and an ecological study of each was considered an urgent necessity. The five facets noted in the last paragraph apply, as they do to the more widespread species, but the latter may not urgently need safeguarding.

Eurychoric species. Eurychoric species have tended to decline through changes in land use, but the species are not threatened *per se* because of the regional nature of the declines. Countering regional extinctions, however, is an urgent and important facet of butterfly conservation, for which two main strategies were seen as important: the safeguarding of selected sites supporting a high diversity of butterfly species, both in meadowland and woodland ecosystems, and the creation of substitute or replacement habitats in suitable areas, outside agricultural land and intensively managed forests, and within the natural dispersal range of butterflies which would be expected to colonize them. Kudrna (1986) also discussed ways of fostering traditional management for butterfly habitats in this category, including possibilities of financial incentives to land owners to leave particular sites in this category.

Incipient species. Incipient species — populations which differ in some way or other from the 'norm' for the species — is a rather loose category, but is important in emphasizing biological diversity. A given population that is taxonomically indistinct from a named and generally recognizable butterfly may, for example, differ in larval food plant or number of generations each year, with such differences being both constant and heritable. They may thus be difficult to detect, and Kudrna (1986) suggested that, once they are detected or suspected, site protection is important, because such butterflies are usually very restricted in distribution.

In the past, many butterfly conservation cases have stopped with reservation of the habitat, with 'reservation' being considered synonymous with 'protection'. Butterflies have played a major part in demonstrating the need for subsequent management for invertebrates, because it has sometimes been very obvious that they became extinct on nature reserves — the case in the UK was discussed by Thomas (1984). Indeed, some species did worse on nature reserves than on unprotected areas, because their precise needs were not understood (Warren, 1992). Early attempts to conserve rare or local butterflies on nature reserves in the UK were often disappointing (Warren, 1992), simply because they were not managed with adequate knowledge of the target species, and some became increasingly isolated as use of surrounding lands changed. Even some reserves set up especially for particular butterflies (such as *Mellicta athalia*, p. 169, and *Maculina arion*, p. 171) were not adequate to prevent extinctions. Warren (1992, 1993a) noted that 'scarce resources were concentrated on reserves acquisition and little was then available for their proper management.'

However, it is *only* in the UK and western Europe, the United States, Japan, Australia and a few other developed areas where the knowledge and expertise are readily available to plan recovery programmes for particular butterfly taxa, and these activities are now becoming a well-defined facet of conservation strategies. Such programmes are expensive and it is highly unlikely that all potential target species could ever be funded for a minimum of three to five years of autecological studies unless far greater finance than ever before was made available. As Murphy *et al.* (1990) stressed, such studies are needed to analyze the viability of the populations involved. But even then, should this be considered the major practical priority? In recent years, several authorities have emphasized that a high proportion of the resources available for butterfly conservation in northern temperate areas has been directed at particular taxa, rather than primarily at conserving diversity or communities of butterflies, and often at taxa which are on the fringe of their geographical range and thereby perhaps unusually prone to extinction there, but *which are common elsewhere*. Several of the UK butterflies of major concern, for example, are ones which are widespread in continental Europe and, whereas their conservation is undoubtedly significant in national terms, it might not be accorded such high priority on a global or continental basis. Indeed, declines of such species in more central parts of their European range — often in central or southern Europe in this context — may be of much greater concern than of range-edge populations (Balletto, 1992; Kudrna, 1995; Munguira, 1995).

However, the creation of a network of reserves in southern Europe (Italy: Balletto and Kudrna, 1985; Balletto, 1992) may be the major need for butterflies in regions subjected to increasing and ever-intensive anthropogenic change. This principle becomes paramount for other parts of the world, where very little detailed knowledge of butterflies is available and species-level focus is largely impracticable other than employing these as umbrella and flagship taxa to promote the broad conservation need (Brown, 1991, 1996; Larsen, 1995).

The kinds of butterfly conservation pursued in temperate regions, and which constitute most of the case histories of Chapter 7, are unlikely to become commonplace in much of the rest of the world — especially in the tropics. Many temperate-region cases involve the restoration or recreation of habitat. In contrast, in the tropics the major emphasis will probably continue to be urging the need to reserve prime undisturbed habitats and the importance of maintaining these in a natural condition — fundamentally, a policy of reservation with minimal intervention as a mode of conserving diversity for at least the moderate-term future. Such areas may be easier to reserve because of the presence of other taxa which can function as 'umbrella species'. A proposed Special Reserve on Madagascar, for example, (Fowler *et al.* 1989) is likely to be established more because of the presence of rare birds and lemurs than for the poorly known forest nymphalid *Apaturiopsis kilura* which also occurs there, and there are a number of parallel cases for continental Africa and south-east Asia. Knowledge of the presence of rare butterflies in such areas can, of course, help to bring additional pressure to bear for reservation and protection, and a high butterfly diversity can be a distinct 'plus' for tourists.

Owen (1971a), in concluding a short chapter on conservation in his *Tropical Butterflies*, commented, 'It must be admitted that prospects for the conservation of butterflies in the tropics are not good'. Because of the needs of burgeoning human populations depending on subsistence agriculture, the 'locking up' of prime areas in nature reserves is viewed as a luxury, and an activity difficult to support unless some tangible (that is, material or financial) benefit can be derived from doing this. Earlier, we noted the possibility of developing local butterfly farming as a cash industry. There is little doubt that, with some safe-guards to control excessive direct collecting of rare species, this kind of operation could be developed in many parts of the tropics if the initial impetus and capital cost can be found. Few people have so far been prepared or able to take local initiatives along these lines, and most of the more successful and self-sustaining operations (as opposed to mere active collecting for sale) have resulted from the efforts of expatriates. There seems little practical prospect of this option becoming a major local saviour of most tropical butterflies, but the need for financial recompense — and the need for fund raising for many aspects of butterfly conservation — is more general. For the European butterflies, Kudrna (1986) suggested voluntary financial contributions from people who gain from commercial activities involving butterflies. Thus, for insect 'sale days' or 'butterfly fairs', a levy could be added to the cost of each specimen and also to admission prices for customers, and advertisers of dead stock and publishers of magazines containing such advertisements could also be asked to contribute; in such cases, donations might well be tax deductible. Similar levy schemes applied to tropical fauna merit serious consideration, assuming they could be promulgated effectively. Tropical butterflies would also be among major beneficiaries of schemes recently suggested on a broader basis, by which countries provide land for nature in payment for some parts of their vast international debts.

Immense and urgent pressure is needed to safeguard the habitats of butterflies in the tropics and to progressively set priorities for this. With our current knowledge of butterfly distributions, some 'centres of diversity' and other critical faunas can indeed be delimited as priorities for reserves. In their detailed appraisal of the Danainae, Ackery and Vane-Wright (1984) selected 31 areas which, if conserved adequately, would include representative populations of all

147 species known. Many of these areas (Figure 9.1) are in the Indo-Pacific region and, whereas conservation in all may not occur, more than half of the Indo-Pacific species (and more than 40% of the world fauna — 69 species) could be conserved by safeguarding the faunas of Sulawesi, Biak, Mindanao and New Guinea. This would be extraordinarily difficult in both political and practical terms. A number of nature reserves already exist in most of the 31 areas, but monitoring and safeguarding these is often not practicable. There are, though, some encouraging signs of effective interest in some places. An international butterfly conference in Indonesia in 1993 led to a resolution calling strongly for conservation of the fauna; groups of workers in India are actively reviewing the status of the many species listed on the Indian *Wildlife Protection Act*; and the Lepidopterological Society of Japan's *Manifesto on Conservation* and its recent update (the 'Osaka Statement': Anon., 1996) calls strongly for a halt to timber extraction from tropical forests because of its devastating effects on butterflies and their habitats.

Brown has long advocated that areas of endemism and diversity for neotropical butterflies, and their peripheries, may be particularly important as Pleistocene refugia areas; many rare primitive butterflies occur *only* irregularly around the edges of such centres. Forty-four strongly marked centres of butterfly subspecies endemism occur in the neotropics. Typically these are areas of 50 000–20 000 square kilometres, which commonly have 15–25 different endemic subspecies. There is substantial overlap between palaeoecological forest refuges and endemism centres for butterflies (Brown, 1987, 1991, 1996).

There is little doubt, therefore, that it is feasible to identify priority faunas and priority areas which need to be safeguarded in many less-developed but

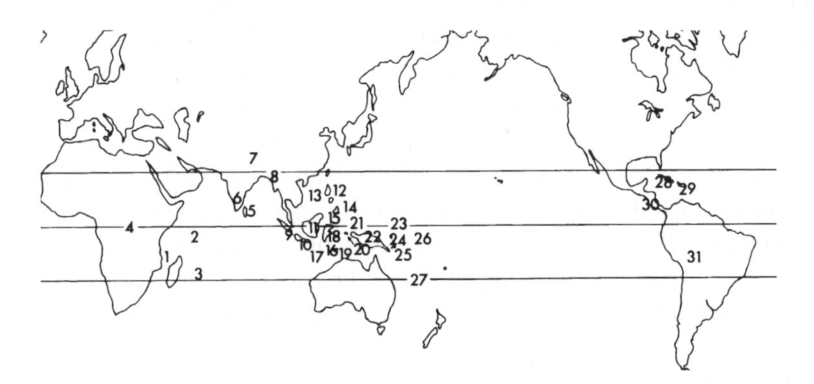

Figure 9.1 Map to illustrate critical regional faunas of milkweed butterflies (Nymphalidae, Danainae), which are regarded as conservation priorities. Regions are: 1) Comoro Islands; 2) Seychelles; 3) Mauritius; 4) Zaire; 5) Sri Lanka; 6) southern India; 7) Nepal; 8) Burma; 9) Sumatra; 10) Java; 11) Borneo; 12) Luzon; 13) Negros; 14) Mindanao; 15) Sulawesi; 16) Sumbawa; 17) Sumba; 18) Flores; 19) Timor; 20) Seram; 21) Biak; 22) New Guinea; 23) New Ireland; 24) New Britain; 25) Guadalcanal; 26) San Cristobal; 27) New Caledonia; 28) Hispaniola; 29) Cuba; 30) Costa Rica; 31) Bolivia (Ackery and Vane-Wright, 1984).

rapidly changing parts of the world. But the approaches to conservation which have moulded much of the philosophy of butterfly conservation prevalent elsewhere are simply not practicable or appropriate there, as Morris (1987) has stressed. Conservation in the 'Third World' is closely linked with utilization of resources, and the conflicting demands on land use attendant on satisfying even minimal living requirements for large human populations ensure that the major thrusts for butterfly conservation apparent in Europe and North America will not occur. As Morris commented, 'On a world scale, insect conservation in Britain is of small significance, apart from the experience and knowledge which can be deployed elsewhere'. It can only be deployed elsewhere in a sympathetic social climate, and the emphasis given to it earlier in this book underlines its vital significance.

But it is perhaps *only* by concentrating on the 'direct' (material or financial) benefits of butterfly conservation that any sympathy for it will supplant short-term exploitation of the environment over much of the world, and the rational sustainable use of butterflies as a commercial resource seems to be one of the more constructive avenues to attempt to pursue (Orsak, 1993). As we have noted, the pioneering aspects of the Papua New Guinea scheme (pp. 129–32) merit serious consideration for adoption elsewhere. Its promotion in the Arfak Mountains of Irian Jaya has recently been suggested, for example (Craven, 1989), to allow (among other things) culturally appropriate use of the land with access to a cash economy through substainable exploitation.

Many tourists now wish to see butterflies in the wild, and increased leisure and wealth in the 'affluent world' may engender tropical butterfly-watching safaris in addition to the more widespread bird-watching and 'wildlife' vacations which have proliferated in recent years. Collecting trips, expeditions organized especially for wealthy collectors wishing to obtain rare species of butterflies, have been more common. There are, for example, reports of organized trips to the Himalaya, with stories of collectors waiting on hilltops for rare species such as *Teinopalpus* swallowtails to arrive and every specimen, in fine or worn condition, being captured indiscriminately (Collins and Morris, 1985). Even though such trips may be damaging in terms of conservation, some may do very little harm to the species *if* their intensity and distribution is regulated in critical areas, perhaps through local agencies. Following, and extending from, the Queensland scheme of collecting 'royalties' for specimens of *Papilis ulysses* and local birdwings captured under permit, there could be some funding raised for conservation activities. In a few cases, rare butterflies have been confiscated by customs or government agencies and it has been suggested, for instance, that a series of *O. alexandrae* taken by Papua New Guinean authorities from people trying to export them illegally could be sold individually (perhaps by a 'sealed bid' method) by collectors to raise a probably substantial sum towards conservation of the species.

There seems to be considerable scope for promotion of 'butterfly safaris' — not necessarily collection-based — and if these were organized properly to visit major conservation areas of scenic grandeur, this could enhance integrity and protective measures of such reserves. The twin strategies of increasing awareness of butterfly conservation and educating people about these remarkable insects together represent a major hope for the wellbeing of tropical butterflies. Responsibly organized trips to the tropics to *view* butterflies are gradually becoming more frequent, as part of a generally increasing interest in ecotourism.

Apart from butterfly farming, establishment of more butterfly exhibits and live displays of other invertebrates in the tropics would undoubtedly be worthwhile. The butterfly house at Penang (West Malaysia) caters for a substantial tourist flow and has clearly also fostered local interest in insects. In South-East Asia especially, other similar displays could be substantial revenue earners and education vehicles. Involvement of local school teachers and tertiary institutions could be fostered, and direct 'hands on' training in butterfly study for park rangers and the like could also be provided. Local benefits can be combined with export and trade of selected livestock and deadstock to encourage both tourist and trade dollars — together a powerful incentive to conservation interest.

BUTTERFLIES IN CONSERVATION ASSESSMENT

As this book implies, butterflies are the most frequent and popular conservation targets among invertebrates. The information available on their ecology, distributions and strategies for survival may well be applicable in broader conservation contexts, and the possible values of butterflies as 'umbrella taxa' and 'ecological indicators' have been noted. They fulfil many of the characteristics generally considered desirable in such categories (Table 9.1) and could, therefore (1) help in the conservation of lesser-known co-occurring biota, and (2) reflect the condition of natural environments by the assemblage composition changing as habitat quality changes in some predictable or definable form, or by the butterflies acting as surrogates for wider estimates of diversity. They could

Table 9.1 Features which render Lepidoptera suitable as an 'umbrella group' in conservation of biodiversity.

1. Abundance.
2. Species richness, with many taxa restricted to particular habitat kinds.
3. Diversity of higher taxonomic groups with different ecological needs.
4. Ease of sampling and sample interpretation.
5. Relatively advanced species-level taxonomy.
6. Response to environmental changes, including:
   (a) vegetation/major habitat,
   (b) microclimate/climate, and
   (c) anthropogenic disturbance.

therefore become a highly important focal group in many practical conservation contexts.

It is not unusual for investigations of rare butterflies to reveal co-existing taxa of at least equivalent conservation interest, which might benefit from being considered in the overall management plan developed for the target butterfly. Thus, Packer's (1991) survey of the two Ontario sites of the Karner Blue (*Lycaeides melissa samuelis*) yielded a bee new to Canada, the second known specimen of another bee, and a rare parasitic wasp, which might not otherwise have been detected.

Umbrella taxa need not be rare, though the current focus of much butterfly conservation activity dictates that they frequently are; it is more important that they are truly representative of a habitat and, incorporating indicator taxon qualities, respond to change in it, as well as being easy to monitor or assess.

In one of very few studies which directly assesses the value of a butterfly as an 'umbrella species', for the severely threatened serpentine soil-based grasslands of central California, Launer and Murphy (1994) demonstrated the value of the Bay Checkerspot butterfly (p. 165) in this context. Conservation activities undertaken for the butterfly would also help to protect other aspects of the unique community. If all the occupied sites were preserved, some 98% of the 133 native spring-flowering plants would also receive some protection. If only the sites supporting the largest butterfly colonies were preserved, or those sites regarded as of only marginal value were lost, the proportion of plants protected dropped markedly, to 74% (98 of 133 species). Because of such 'leakiness' in the 'umbrella', Launer and Murphy suggested the use of multiple species to increase the overall effect.

As was the case at Mount Piper in Victoria (p. 181) (Britton *et al.*, 1995), parallel studies of several distinct species with somewhat different ecological needs, but which live together in the same specialized habitat, may increase the overall umbrella effect, simply because any form of ecological complementarity between the targeted species may help to conserve a wider resource base. In both these studies, diurnal moths were investigated in conjunction with rare butterflies.

Even without detailed ecological knowledge single species which are adopted as 'flagship' species in gaining widespread sympathy for their conservation, may also be made effective umbrellas. The recent international conservation effort to conserve Queen Alexandra's Birdwing in Papua New Guinea (p. 155) is pursuing its role in reducing the destruction of primary forest, on which myriad lesser-known organisms depend, through increasing returns to local people based on secondary habitats and thus reducing their immediate need to clear more forest for subsistence agriculture.

The general theme of 'ecological indicators' is of immense importance in invertebrate conservation (New, 1995), and for butterflies it is a central facet of using them in conservation planning. In parts of the western United States, butterflies are at least as useful as vertebrates (Murphy and Wilcox, 1986) and are less sensitive to the constraints of smaller areas, and indeed are better for recognition of habitats and plant communities. Different components of butterfly assemblages usually differ in their use in indicator contexts. For the Neotropics, Brown (1991) assessed the relative likely value of several groups of Lepidoptera as indicators. Differences between less suitable and more suitable taxa reflected, in particular, relativity of ecological specialization, with greater specialization (as being site-restricted, relatively sedentary, narrowly endemic, easily obtainable as large random samples, or associated with and indicative of other species and specific resources) reflecting greater indicator use. Thus Ithomiine species richness seems to have a strong positive relationship with overall butterfly richness in mainland Central and South America (Beccaloni and Gaston, 1995), in which case they can be valid surrogates or 'predictors' of such wider diversity.

As more specific contexts of Neotropical butterflies having indicator value, Brown (1991) noted:

- large species or those with special rare resources, which disappear from small areas after isolation (some Nymphalidae: Morphinae, Brassolinae, Charaxinae);
- small shade-loving species with uniform (Satyrinae) or very patchy (Ithomiinae) forest distributions;
- sun-loving Nymphalidae, such as Heliconiini, are good indicators of edge effects, in that they appear, multiply and crash in relation to fresh plant growth on newly created edges or clearings; and
- Lycaenidae, because of their great ecological diversity, can variously indicate all possible stages of degradation.

The Neotropical Ithomiinae situation noted above is similar to that of a diverse satyrine genus, *Henotesia*, in Madagascar (Kremen, 1994). An inventory of butterflies in the eastern forest region could be limited to this genus (which contains 42 species, and is one of the richest genera in the country, with 97.6% endemism), and retain significant information on environmental patterns, including an indication of disturbance gradients. A more speciose second genus of Satyrinae, *Strabena* (55 species, wholly endemic) was not as informative, because of lower abundance and greater uniformity. Kremen (1994) made the important point that the selection of such target groups to help in providing good quality inventory information is likely to be region-dependent, and that distributional and ecological knowledge may need to be acquired to test criteria such as:

- a minimum number of species being present in the study area for the target group to successfully represent the higher taxon;

- at least some species should be abundant in the study area;
- distributions of members of the assemblage should collectively encompass the entire study area (that is, all elevations and all major habitat types); or
- species assemblages with high beta or gamma diversity should be preferred over entirely uniform groups restricted in their selection of habitats, or entirely generalist groups.

There seems little doubt that, at least in some situations, Lepidoptera can be useful indicators of habitat condition or change, not least because of the total or near-total dependence of many species on particular successional stages or vegetation condition, of the kinds discussed elsewhere in this book. Indeed, much of butterfly conservation management specifically targets such parameters. The spectrum of species present in European grasslands, for example, differs substantially in relation to successional stage and cultivation regimes (Erhardt, 1985; Erhardt and Thomas, 1991), with a general correlation of Lepidoptera diversity with plant diversity. Related contexts involve the use of butterflies to indicate or monitor the success of broader restoration efforts, such as the re-establishment of riparian sites (Nelson and Andersen, 1994) and surface-mining reclamation (Holl, 1996). Such studies have tended to reveal a spectrum of sensitive to resilient butterfly species, in relation to particular kinds of vegetation; as Nelson and Andersen (1994) noted, species susceptible to disturbance should still be present in undisturbed reference sites, and disturbed areas are characterized by loss of such 'specialist' species and a predominance of more tolerant 'generalist' forms.

Although the relatively low species diversity in many temperate region faunas renders such studies reliable, the complexity of tropical butterfly assemblages and their habitats can pose problems for reliable sampling (Sparrow *et al.*, 1994).

There is little doubt that butterflies will continue to spearhead much work associated with conservation of terrestrial invertebrates, and to maintain or elicit concern from many people who would otherwise not participate in such activities. Although the close association of butterflies with plants can sometimes reduce their value as ecological indicators — it is easier to find and recognize persistent plants than the transient moving insects which feed on them — the dramatic responses by many butterflies to environmental changes not linked merely with presence or absence of particular plants but with very subtle changes in their status, conspicuousness or dispersion, indicates strongly that many other invertebrates are likely to respond in similar ways. Much recent conservation site assessment has emphasized the need to clarify and quantify the various criteria (rarity, 'typicalness' and others) used commonly in this. Usher (1980) suggested that there was a 'paradox of rarity and typicalness', as sites with many rare species could not be typical and so 'typical' sites did not contain many rare species. This has been queried by Eyre and Rushton (1989), who quantified site parameters for these qualities using water beetles in the UK, and found no consistent relationship. However, both criteria clearly apply to the needs of butterflies for a high level of effective conservation. A large number of rare species can represent a 'critical fauna' and give a site (or larger scale geographical area) preferred ranking over one where fewer rare species occur. Detection of such areas, as well as those of high diversity and 'typicalness' in representing undisturbed natural communities, irrespective of the presence of rare butterfly species, will continue to be a major aim of documenting butterfly faunas and mapping their distributions. They match the principles of site reservation in Europe noted by Kudrna (1986) — diversity of nominal species and the presence of highly valued species on sites — but the two are not necessarily needed for the *same* sites. Mapping distributions, now well organized in parts of Europe in particular (but still without effective regional coordination, although a current project organized by Dr O. Kudrna is seeking to redress this), is logistically intensive, although there is considerable potential for voluntary assistance in many aspects of this.

Valuable publicity can be gained in regions where temptations to collect are minimal by making available guides to people wishing to see butterflies: a recent site guide for UK butterflies and dragonflies (Hill and Twist, 1996) gives detailed directions for visitors to about 75 key sites, with details of access, notable species present, their seasonality and other biological features, and may prove important in fostering goodwill — in contrast to 'locking up' such sites to reduce public access and appreciation.

Perhaps the major need, though, is for wider and more positive acceptance of the *desirability* of conserving butterflies. One of the great pioneers of butterfly conservation, J. Heath, wrote, 'Without the acceptance of this by politicians, legislators, planners as well as the public . . . all other efforts are likely to be at most of only short term benefit to the Rhopalocera. The entomological community in particular must assume a major role in bringing the problems of insect conservation to the notice of both the authorities and the public' (Heath, 1981). At least some of the problems of butterfly conservation have solutions on the horizon, but for many species (and their dwindling habitats) the prospects cannot be considered good until, and unless, this vital communication gap is bridged. Many butterflies are likely to have become extinct by the time this occurs.

APPENDIX 1

IUCN STATUS CATEGORIES

Two sets of IUCN categories are noted below. The first has been traditionally used and is given here to facilitate the interpretation of status categories given in the text. The second (adopted in 1994) gives greater emphasis to objective numerical appraisal, and has not yet been applied widely to butterflies. It may be very difficult to allocate a butterfly to one or other of these confidently and, in general, conservation importance decreases as we move down the list.

Pre-1994 IUCN Categories

1. **Extinct** Taxa not definitely located in the wild during the past 50 years.
2. **Endangered** Taxa in danger of extinction, and whose survival is unlikely if the causal factors continue operating. This includes taxa which are in immediate danger because of habitat destruction and those which may be already extinct but which have been seen during the last 50 years.
3. **Vulnerable** Taxa believed likely to become endangered in the near future if causal factors continue operating.
4. **Rare** Taxa with small world populations which are not at present 'Endangered' or 'Vulnerable', but are at risk.
5. **Indeterminate** Taxa known to be in one or other of categories two to four but for which there is insufficient information to determine which.
6. **Out of Danger** Taxa formerly included in one of 2–5 above but which are now considered relatively secure because of effective conservation measures.

The term 'Threatened' is a general one to denote taxa in any of categories two to five.

Categories adopted in November 1994

Extinct (EX) A taxon is Extinct when there is no reasonable doubt that the last individual has died.

Extinct in the Wild (EW) A taxon is Extinct in the wild when it is known only to survive in cultivation, in captivity or as a naturalized population (or populations) well outside the past range. A taxon is presumed extinct in the wild when exhaustive surveys in known and/or expected habitat, at appropriate times (diurnal, seasonal, annual), throughout its historic range have failed to record an individual. Surveys should be over a time frame appropriate to the taxon's life cycle and life form.

Critically Endangered (CR) A taxon is Critically Endangered when it is facing an extremely high risk of extinction in the wild in the immediate future as defined by any of the criteria A–E:

A. Population reduction in the form of either of the following:
   1. An observed, estimated, inferred or suspected reduction of at least 80% over the last 10 years or three generations, whichever is the longer, based on (and specifying) any of the following:
      (a) direct observation
      (b) an index of abundance appropriate to the taxon
      (c) a decline in area of occupancy, extent of occurrence and/or quality of habitat
      (d) actual or potential levels of exploitation
      (e) the effects of introduced taxa, hybridization, pathogens, pollutants, competitors or parasites
   2. A reduction of at least 80%, projected or suspected to be met within the next ten years or three generations, whichever is the longer, based on (and specifying) any of (b), (c), (d) or (e) above.

B. Extent of occurrence estimated to be less than 100 km^2 or area of occupancy estimated to be less than 10 km^2, and estimates indicating any two of the following:
   1. Severely fragmented or known to exist at only a single location.
   2. Continued decline, observed, inferred or projected, in any of the following:
      (a) extent of occurrence
      (b) area of occupancy
      (c) area, extent and/or quality of habitat
      (d) number of locations or subpopulations
      (e) number of mature individuals.
   3. Extreme fluctuations in any of the following:
      (a) extent of occurrence
      (b) area of occupancy

(c) number of locations or subpopulations
(d) number of mature individuals.

C. Population estimated to number less than 250 mature individuals and either:
   1. An estimated continuing decline of at least 25% within 3 years or one generation, whichever is the longer, or
   2. A continuing decline, observed, projected or inferred, in numbers of mature individuals and population structure in the form of either:
      (a) severely fragmented (i.e. no subpopulation estimated to contain more than 50 mature individuals), or
      (b) all individuals are in a single subpopulation.

D. Population estimated to contain less than 50 mature individuals.

E. Quantitative analysis showing the probability of extinction in the wild is at least 50% within 10 years or 3 generations, whichever is the longer.

Endangered (EN) A taxon is endangered when it is not Critically Endangered but is facing a very high risk of extinction in the wild in the near future, as defined by any of the criteria A–E:

A. Population reduction in the form of either of the following:
   1. As for CR, but reduction of at least 50%
   2. As for CR, but reduction of at least 50%.

B. Extent of occurrence estimated to be less than 5000 km^2 or area of occupancy estimated to be less than 500 km^2, and estimates indicating any two of the following:
   1. Severely fragmented or known to exist at no more than five locations.
   2, 3. As for CR.

C. Population estimated to number less than 2500 mature individuals and either:
   1. As for CR, but decline of at least 20% within 5 years or 2 generations
   2. As for CR but (a) no subpopulation estimated to contain more than 250 mature individuals.

D. Population estimated to number less than 250 mature individuals.

E. As for CR, but probability of at least 20% within 20 years or 5 generations.

Vulnerable (VU) A taxon is Vulnerable when it is not Critically Endangered or Endangered but is facing a high risk of extinction in the wild in the medium-term future, as defined by any of the criteria A–E:

A. Population reduction in the form of either of the following:
   1. As for CR, but reduction of at least 20%
   2. As for CR, but reduction of at least 20%

B. Extent of occurrence estimated to be less than 20 000 km^2 or area of occupancy estimated to be less than 2000 km^2, and estimates indicating any two of the following:

1. Severely fragmented or known to exist at no more than ten locations
2, 3. As for CR.

C. Population estimated to number less than 10 000 mature individuals and either:
   1. As for CR, but decline of at least 10% within 10 years or 3 generations
   2. As for CR but (a) no subpopulation estimated to contain more than 1000 mature individuals.
D. Population very small or restricted in the form of either of the following:
   1. Population estimated to number less than 1000 mature individuals.
   2. Population is characterized by an acute restriction in its area of occupancy (typically less than 100 km^2) or in number of locations (typically less than 5).
E. As for CR, but probability of at least 10% within 100 years.

Lower Risk (LR) A taxon is Lower Risk when it has been evaluated, and does not satisfy the criteria for any of the above categories. Taxa included can be separated into three subcategories:

1. Conservation Dependent (cd) Taxa which are the focus of a continuing taxon-specific or habitat-specific conservation programme targeted towards the taxon in question, the cessation of which would result in the taxon qualifying for one of the threatened categories above within a period of five years.
2. Near Threatened (nt) Taxa which do not qualify for c.d., but which are close to qualifying for VU.
3. Least Concern (lc) Taxa which do not qualify for c.d. or n.t.

Data Deficient (DD) A taxon is Data Deficient when there is inadequate information to make a direct, or indirect, assessment of its risk of extinction based on its distribution and/or distribution. DD is therefore not a category of threat or Lower Risk: listing of taxa in this category indicates that more information is required and acknowledges the possibility that future research will show that threatened classification is appropriate.
Not Evaluated (NE) A taxon is Not Evaluated when it has not yet been assessed against the criteria.

APPENDIX 2

POINTS FOR CONSIDERATION IN DEVELOPING A SPECIES RESTORATION STRATEGY FOR BUTTERFLIES IN THE UK

1. The species should have declined seriously (or be threatened with extinction) at a national or regional level.
2. Remaining natural populations should be being conserved effectively, and the restoration plan should be an integral part of a Species Action Plan.
3. The habitat requirements of the species and the reasons for its decline should be broadly known and the cause of extinction on the receptor site (where re-introduction is contemplated) should have been removed. There should be a long-term management plan which will maintain suitable habitat, and the site should be large enough to support a viable population in the medium to long term.
4. Extinction should have been confirmed at the receptor site (at least 5 years recorded absence), the mobility of the target species should be assessed and natural re-establishment should be shown to be unlikely over the next 10–20 years.
5. Opportunities to restore networks of populations or metapopulations are preferable to single site re-introductions (unless the latter is a necessary prelude to the former).
6. Sufficient numbers of individuals should be used in the re-introduction to ensure a reasonable chance of establishing a genetically diverse population.
7. As far as possible the donor stock should be the closest relatives of the original population, and genetic studies should be carried out where doubt exists.

8. The receptor site should be within the recorded historical range of the species.
9. Removal of livestock should not harm the donor population (donor populations may have to be monitored during the re-introduction programme).
10. The re-introduction should not adversely affect other species on the site.
11. If captive bred livestock is used, it should be healthy and genetically diverse (e.g. not normally captive bred for more than two generations).
12. Re-introduced populations should be monitored for at least five years, and contingency plans should be made in case the re-introduction fails, the donor population is adversely affected, or other species are adversely affected.
13. Approval should be obtained from the Conservation Committee of Butterfly Conservation and all other relevant conservation bodies and organizations (including statutory bodies in the case of scheduled species, SSSIs, etc.).
14. Approval must be obtained from the owners of both receptor and donor sites.
15. The entire process should be fully documented and standard record forms completed for Butterfly Conservation and JCCBI.

(Source: Butterfly Conservation, 1995)

APPENDIX 3

AN ARTIFICIAL DIET FOR REARING BUTTERFLY LARVAE (FROM MORTON 1981)

Solids	% Composition (w/w)
Soy flour	7
Wheatgerm	6
Yeast extract	6
Sucrose	3.6
Dried plant material*	1.5
Wesson salts	1
Ascorbic acid	0.4
Potassium sorbate	0.2
Methyl parahydroxybenzoate	0.15
Aureomycin (veterinary grade)	0.023
Agar	1.9
Liquids	
Formaldehyde solution (10%)	0.43
Potassium hydroxide (4M)	0.8
Acetic acid (25%)	1.14
Choline chloride solution (50%)	0.23
Distilled water	to mass

* washed, oven dried at 110°C until powders easily.

Stir agar into approx. 60% of distilled water, boil for 3–10 minutes while stirring. Mix remaining solids thoroughly in rest of distilled water, then add cooled agar and blend. Then add liquids in sequence, with two minutes blending between each. Dispense diet to containers and leave to dry slightly for 24 hours.

GLOSSARY

abdomen The posterior body region (tagma) of an insect. In the adult butterfly, this contains the reproductive organs and genitalia at the tip. The caterpillar abdomen has several pairs of prolegs, including a posterior pair of strongly developed claspers.

aestivation A period of dormancy as a regular feature of the life cycle during a dry season or summer. See **diapause**.

bivoltine Having two generations each year — commonly in butterflies these are a 'spring generation' and a 'summer generation', or a 'wet season generation' and a 'dry season generation'.

body basking A posture in which the wings are opened sufficiently to expose the dorsal body surface to sunlight as a method of increasing body temperature, or thermoregulation.

chrysalis The pupa, or developmental stage between the larva (caterpillar) and adult butterfly. 'Chrysalis' tends to be applied only to butterflies, as a special use of 'pupa', applicable to all endopterygote insects.

co-evolution The evolution of two or more species in relation to each other. Often used to describe feeding relationships where, in butterflies, the butterfly species may respond to changes in its food plant by adaptation, and the plant correspondingly responds to changes in the consumer species.

conspecific Belonging to the same species; can be used as an adjective or a noun.

crypsis The condition of being cryptic or hidden, usually used for butterflies in the sense of camouflage markings or patterns which enable the insect to resemble its background when at rest.

diapause A form of dormancy, characteristic of a particular stage of an insect's life cycle, during which metabolic activity is greatly decreased during a certain (inclement) period of the year.

dimorphism The state of having two distinct forms or morphs within a single species, for example male and female butterflies often differ in appearance, constituting sexual dimorphism, and individuals may differ at different times of the year, constituting seasonal dimorphism.

diversity A term widely used in ecology and with several different meanings. Simply, diversity can mean the number of species present, but diversity also generally includes estimation of the numerical equivalence or equitability of those species. Diversity is 'high' if there are many species and they are of similar abundance, and is low if there are few species with uneven abundances. A 'diversity index' expresses, in various ways, the relationship between the number of species and the numbers of individuals.

dorsal basking A posture in which the wings are fully open to maximize incidence of sunlight and increase body temperature.

generation time The period of a life cycle, generally expressed as the time from adult to adult via egg, larva and pupa.

head The anterior body region (tagma) of an insect. In the adult butterfly it incorporates sensory structures (antennae, eyes) and the suctorial mouth parts (proboscis) and labial palpi. The caterpillar head, in the form of a hardened capsule, has a group of simple eyes (stemmata), small antennae and chewing (mandibulate) mouth parts.

hill-topping The habit of some butterflies, especially males, of seeking out and assembling at high points to facilitate finding mates.

inbreeding Sexual reproduction within the same gene pool or population; that is, mating between siblings or close relatives with genes in common.

indicator species Species characteristic of particular climate, soil, or (for butterflies) vegetational regimes and which reflect the presence of particular ecological conditions. If these conditions change the species concerned may respond by changing its abundance, thus reflecting the 'health' of the community or habitat.

instar A stage in development between successive moults in an insect larva. On hatching from the egg a caterpillar is referred to as a 'first instar'; after moulting for the first time, it becomes a 'second instar', and so on.

lateral basking A posture in which the wings are closed and the body inclined to present the ventral side of the wings to sunlight.

management In conservation, the practice of changing an ecosystem or habitat deliberately in order to preserve or augment particular environmental features needed by given species.

mark-release-recapture A method of population size estimation based on the principle of marking live individuals, releasing them unharmed into the natural population and, at a later time, capturing a number of individuals and assessing population size from the ratio of marked to unmarked individuals.

metamorphosis Literally 'change of form', referring to the transformation from larva to adult in the insect life cycle. Many insects have an 'incomplete metamorphosis' or gradual transition from larva to adult. Others, including butterflies, have a 'complete metamorphosis', in which the larva and adult, very different in form, are separated by a pupa.

metapopulation A form of population structure whereby a number of demographic units exist on patches of habitat over a site. These may become extinct, but the general persistence of the species in the area is maintained by colonization of vacant or vacated habitat patches to counter local extinctions. In effect, the species persists by rolling extinction/colonization cycles, sometimes based on large, permanent 'source' populations.

migration Change of habitat by dispersal, usually involving the whole, or a significant part, of the population. It can be related to climatic factors, season, or food supply and contrasts with 'trivial dispersal', which refers to individual movements without any change of habitat. Adaptive dispersal.

mimicry The resemblance between animals of different species so that one or more gain protection from predation. In Batesian mimicry a harmless species (the mimic) resembles a toxic or distasteful one (the model) which is often conspicuously marked, and predators tend to avoid both if they has learned to associate the appearance with unpalatability by attacking the model. In Mullerian mimicry, a number of conspicuous distasteful species resemble each other and all gain protection, as predators learn to avoid all individuals with a similar colour pattern or markings after attacking just one of these.

monophagy The condition of being able to feed only on one kind of food — in butterflies, normally applied to species which can feed as caterpillars on only one species of food plant, that is, specialist feeders.

morph A particular definable colour pattern (in butterflies) occurring in the population. In a monomorphic population, all individuals have the same general appearance; in a polymorphic population a number of different morphs are present. See **dimorphism**.

myrmecophily Association with ants. Applied particularly to many members of the Lycaenidae, whose caterpillars live in ant nests for all or part of their development in some form of mutualistic relationship which is usually highly specific.

natural enemy A general term applied to predators, parasitoids, parasites and diseases; the biotic agents inflicting mortality on a species or population.

oligophagy The condition of feeding on only a few kinds of food — in butterflies, commonly applied to species with caterpillars feeding on several related species of food plants.

oviposition Egg-laying.

phenology The study of timing of periodic natural events. Often applied to the pattern of seasonal development of a species in relation to climatic or other variations.

polyphenism A form of seasonal polymorphism in which the appearance of distinct morphs at different times of the year is influenced by seasonal changes in photoperiod, temperature or humidity rather than any genetic change.

population A group of individuals of a species in a given area or habitat. Usually implies a breeding unit discrete from other populations.

protandry Appearance of males before females, or prior sexual maturation of males. In butterflies, most commonly applied to the former situation, so that apparent sex ratio changes during the flight period of the species.

pupa The developmental stage between the larva and adult in an endopterygote insect; the intermediate transformational stage in a complete metamorphosis. See **chrysalis**.

sex ratio The proportion of males in a population; the ratio of males to females.

speciation The evolution or formation of species; the process by which new species come into existence. Allopatric speciation occurs when segregates are separated in space; sympatric speciation occurs when divergence occurs in a single population without geographical barriers, to produce species or subspecies overlapping or coinciding in distribution.

subspecies A distinctive segregate of a species; usually a geographically defined population reproductively isolated from others of the same species but not considered sufficiently divergent to constitute a distinct species.

succession The natural progression of vegetation by which one community is replaced by others over time. Thus, for example, bare ground is rapidly colonized by herbaceous plants, these in turn may be replaced by shrubs and eventually by larger trees to reach a persistent climax vegetation type such as forest. Early stages in a succession may be passed through rapidly and management (q.v.) may be needed to ensure that they persist if needed by butterflies.

territory An area, or habitat, defended by an individual against intrusion by other organisms, mainly members of the same species. In butterflies, this usually means an area defended by a male as a mating and feeding site.

thermoregulation The process of regulating body temperature actively (by particular postures or habits) or passively, applied to cold-blooded animals which depend on sunlight for raising body temperature in order to pursue normal activities. Both heating and cooling (or avoidance of very high temperatures) may be involved.

thorax The second, middle, body region (tagma) of an insect. In the butterfly this supports three (sometimes two) pairs of legs ventrally, and two pairs of scale-covered wings dorsally.

transect A line or strip (or 'belt') across a habitat used as a unit for studying or assessing abundance of organisms, that is as a sampling base.

univoltine Having one complete generation each year.

warning colouration Bright colours or conspicuous banding patterns on wings or body of butterflies which serve to 'advertise' to possible predators that the bearer is unpalatable or toxic.

BIBLIOGRAPHY

Ackery, P. R. 'Systematics and faunistic studies on butterflies'. *The Biology of Butterflies*. Vane-Wright, R. I. & Ackery, P. R. (eds). London, Academic Press, 1984, 9–21.

Ackery, P. R. & Vane-Wright, R. I. *Milkweed Butterflies, their cladistics and biology*. London, British Museum (Natural History), 1984.

Adams, M. 'Ecological zonation and the butterflies of the Sierra Nevada de Santa Marta, Colombia (Lep.)'. *J. nat. Hist.* 7. 1973, 699–718.

Alonso-Mejia, A., Glendinning, J. I. & Brower, L. P. 'The influence of temperature on crawling, shivering and flying in overwintering monarch butterflies in Mexico'. *Biology and conservation of the Monarch butterfly*. Malcolm, S. B. & Zalucki, M. P. (eds). Los Angeles, Natural History Museum of Los Angeles County, 1993, 309–14.

Anderson, E. & Spry, F. P. *Victorian butterflies and how to collect them*. Melbourne, Hearne, 1893–4.

Andow, D. A., Baker, R. J. & Lane, C. P. *Karner Blue butterfly: a symbol of a vanishing landscape*. Minnesota Agricultural Experiment Station Miscellaneous Publication, 1994.

Anderson, J. B. & Brower, L. P. 'Freeze-protection of overwintering monarch butterflies in Mexico: critical role of the forest as a blanket and an umbrella'. *Ecol. Entomol.* **21**. 1996, 107–16.

Anon. 'Meeting on the mortality of the Monarch butterfly in Mexico'. Avándaro, Mexico, 1993.

Anon. 'The Osaka Statement'. *Decline and conservation of butterflies in Japan*. **III**. 1996, VII–VIII.

Arms, K., Feeny, P. & Lederhouse, R. C. 'Sodium: stimulus for puddling behaviour by tiger swallowtail butterflies, *Papilio glaucus*'. *Science* **185**. 1974, 372–4.

Arnold, R. A. 'Ecological studies of six endangered butterflies (Lepidoptera: Lycaenidae): island biogeography, patch dynamics and the design of habitat preserves'. *Univ. Calif. Publns Entomol.* **99**. 1983, 1–161.

Arnold, R. A. 'Conservation and management of the endangered Smith's Blue butterfly, *Euphilotes enoptes smithi* (Lepidoptera: Lycaenidae)'. *J. Res. Lepid.* **22**. 1983, 135–53.

Arnold, R. A. 'Decline of the endangered Palos Verdes blue butterfly in California'. *Biol. Conserv.* **40**. 1986, 203–17.

Arnold, R. A. & Powell, J. A. '*Apodemia mormo langei*'. *Univ. Calif. Publns. Entomol.* **99**. 1983, 99–128.

Asher, J. (ed.) *A programme for the coordination of butterfly recording in Britain and Ireland.* Cambridge, Butterfly Conservation, 1992.

Aubert, J., Descimon, H. & Michel, F. 'Population biology and conservation of the Corsican swallowtail butterfly *Papilio hospiton* Géné'. *Biol. Conserv.* **78**. 1996, 247–55.

Baker, R. R. 'A possible method of evolution of the migratory habit in butterflies' *Phil. Trans. R. Soc. Lond. B* **253**. 1968, 309–41.

Baker, R. R. 'Bird predation as a selective pressure on the immature stages of the cabbage butterflies, *Pieris rapae* and *P. brassicae*'. *J. Zool.* **162**. 1970, 43–59.

Baker, R. R. 'Territorial behaviour of the nymphalid butterflies, *Aglais urticae* and *Inachis io*'. *J. Anim. Ecol.* **41**. 1972, 453–69.

Balletto, E. 'Butterflies in Italy: status, problems and prospects'. *Future of butterflies in Europe.* Pavlicek-van Beek, T. Ovaa, A. H. & van der Made, J. G. (eds). Wageningen, Agricultural University, 1992. 53–64.

Balletto, E. & Kudrna, O. 'Some aspects of the conservation of butterflies (Lepidoptera: Papilionidae) in Italy, with recommendations for the future strategy'. *Boll. Soc. entomol. Ital.* **117**. 1985, 39–59.

Bandai, K. 'Conservation of the giant purple, *Sasakia charonda* in Nagasaka-cho, Yamanashi Prefecture'. *Decline and conservation of butterflies in Japan.* **III**. 1996, 180–4.

Barnett, L. K. & Warren, M. S. (comp.) 'Species Action Plan. Large Blue, *Maculinea arion*'. Colchester, Butterfly Conservation Joint Committee for the Reestablishment of the Large Blue Butterfly, 1995a.

Barnett, L. K. & Warren, M. S. (comp.) 'Species Action Plan. The Swallowtail, *Papilio machaon*'. Wareham, Butterfly Conservation. 1995b.

Barnett, L. K. & Warren, M. S. (comp.) 'Species Action Plan. Silver-spotted skipper, *Hesperia comma*'. Wareham, Butterfly Conservation, 1995c.

Barnett, L. K. & Warren, M. S. (comp.) 'Species Action Plan. Large Copper, *Lycaena dispar*'. Wareham, Butterfly Conservation, 1995d.

Bates, H. W. *The naturalist on the river Amazons* (2 ed.). London, 1864.

Baughman, J. F. & Murphy, D. D. 'What constitutes a hill to a hill-topping butterfly?' *Am. Midl. Nat.* **120**. 1988, 441–3.

Beccaloni, G. W. & Gaston, K. J. 'Predicting the species richness of neotropical forest butterflies: Ithomiinae (Lepidoptera: Nymphalidae) as indicators. *Biol. Conserv.* **71**. 1995, 77–86.

Begon, M. *Investigating animal abundance: capture-recapture for biologists.* London, Edward Arnold, 1979.

Beirne, B. P. 'Natural fluctuations in the abundance of British Lepidoptera'. *Ent. Gazette* **6**. 1955, 21–52.

Benson, W. W., Haddad, C. F. B. & Zikan, M. 'Territorial behaviour and dominance in some heliconiine butterflies (Nymphalidae)'. *J. Lepidopt. Soc.* **43**. 1989, 33–9.

Bitzer, R. J. & Shaw, K. C. 'Territorial behaviour of the Red Admiral, *Vanessa atalanta* (L.) (Lepidoptera: Nymphalidae)'. *J. Res. Lepid.* **18**. 1980, 36–49.

Bingley, W. *Animal Biography; or, authentic anecdotes of the lives, manners and economy of the Animal Creation, arranged according to the system of Linnaeus.* (2 ed.), 3 vols. London, Richard Phillips, 1804.

Blair, R. B. & Launer, A. E. 'Butterfly diversity and human land use: species assemblages along an urban gradient'. *Biol. Conserv.* **80**. 1997, 113–25.

Blau, W. S. 'The effect of environmental disturbance on a tropical butterfly population'. *Ecology* **61**. 1980, 1005–12.

Boden, R. W. & Ovington, J. D. National planning for conservation areas. *A National System of ecological reserves in Australia.* Canberra, Australian Academy of Sciences, 1975, 86–95.

Booth, M. & Allen, M. 'Butterfly garden design'. *Butterfly gardening.* Xerces Society/Smithsonian Institution (eds). San Francisco, Sierra Club/Washington, National Wildlife Federation, 1990, 69–93.

Bowden, S. R. 'Taxonomy for a variable butterfly?' *Ent. Gazette* **36**. 1985, 85–90.

Braby, M. F. 'The life history and biology of *Paralucia pyrodiscus lucida* Crosby (Lepidoptera: Lycaenidae)'. *J. Aust. ent. Soc.* **29**. 1990, 41–50.

Braby, M. F, Crosby, D. F. & Vaughan, P. J. 'Distribution and range reduction in Victoria of the Eltham Copper butterfly, *Paralucia pyrodiscus lucida* Crosby'. *Vict. Nat.* **109**. 1992, 154–61.

Braby, M. F., van Praagh, B. D. & New, T. R. 'The Dull Copper, *Paralucia pyrodiscus* (Lycaenidae)'. *Biology of Australian Butterflies.* Kitching, R. L. & Jones, R. E. (eds). Melbourne, CSIRO (in press).

Brakefield, P. M. & Larsen, T. B. 'The evolutionary significance of dry and wet season forms in some tropical butterflies'. *Biol. J. Linn. Soc.* **22**. 1984, 1–12.

Brewer, J. *Butterfly Gardening.* Xerces Society self-help sheet No. 7. 1982.

Bright, P. M. & Leeds, H. A. *A monograph of the British aberrations of the Chalk-Hill Blue, Lysandra coridon (Poda 1761).* Bournemouth, Richmond Hill, 1938.

Bristow, C. R. 'The occurrence of *Opsiphanes tamarindi* Felder & Felder (Lepidoptera: Satyridae) in Britain'. *Ent. Rec. J. Var.* **98**. 1986, 96–7.

Britten, H. B., Brussard, P. F. & Murphy, D. D. 'The pending extinction of the Uncompahgre Fritillary butterfly'. *Conserv. Biol.* **8**. 1994, 86–94.

Britton, D. R. 'Ant trap nests enable detection of a rare and localized butterfly, *Acrodipsas myrmecophila* (Waterhouse & Lyell) (Lepidoptera: Lycaenidae) in the field'. *Mem. Mus. Vic.* **56**. 1997, 383–7.

Britton, D. R., New, T. R. & Jelinek, A. 'Rare Lepidoptera at Mount Piper, Victoria — the role of a threatened butterfly community in advancing understanding of insect conservation. *J. Lepidopt. Soc.* **49**. 1995, 97–113.

Brower, A. V. Z. Update on conservation from MONCOM II: Notes on the second international conference on the monarch butterfly and the preservation of overwintering colonies. *Atala* **14**. 1986, 12–14.

Brower, L. P. 'Speciation in butterflies of the *Papilio glaucus* group. I, II'. *Evolution* **13**. 1959, 40–63, 212–28.

Brower, L. P. 'Understanding and misunderstanding the migration of the Monarch butterfly (Nymphalidae) in North America: 1857–1995'. *J. Lepidopt. Soc.* **49**. 1995, 304–85.

Brower, L. P. 'Forest thinning increases monarch butterfly mortality by altering the microclimate of the overwintering sites in Mexico. *Decline and conservation of butterflies in Japan,* **III**. 1996, 33–44.

Brower, L. P., Brower, J. V. & Cranston, F. P. 'Courtship behaviour of the queen butterfly, *Danaus gilippus berenice*'. *Zoologica, N.Y.* **50**. 1965, 1–39.

Brower, L. P. & Malcolm, S. B. 'Endangered phenomena'. *Wings* **14**. 1989, 3–10.

Brower, L. P. et al. 'On the dangers of interpopulational transfers of Monarch butterflies'. *BioScience* **45**. 1995, 540–4.

Brown, K. S., Jr. 'Maximizing daily butterfly counts'. *J. Lepidopt. Soc.* **26**. 1972a 183–96.

Brown, K. S., Jr. 'The heliconians of Brazil (Lepidoptera: Nymphalidae). Part III. Ecology and biology of *Heliconius nattereri*, a key primitive species near extinction, and comments on the evolutionary development of *Heliconius* and *Eueides*'. *Zoologica*, N.Y. **57**. 1972b, 41–69.

Brown, K. S., Jr. 'Geographical patterns of evolution in Neotropical Lepidoptera: systematics and derivation of known and new Heliconiini (Nymphalidae: Nymphalinae)'. *J. Entomol. B* **44**, 1976, 201–42.

Brown, K. S., Jr. 'Biogeography and evolution of neotropical butterflies'. *Biogeography and Quaternary History in Tropical America.* Whitmore, T. C. & Prance, G. T. (eds). Oxford, Clarendon, 1987, 66–104.

Brown, K. S. 'Conservation of Neotropical environments: insects as indicators'. *The conservation of insects and their habitats.* Collins, N. M. & Thomas, J. A. (eds). London, Academic Press, 1991, 350–404.

Brown, K. S. 'Conservation of threatened species of Brazilian butterflies'. *Decline and conservation of butterflies in Japan.* **III**. 1995, 45–62.

Brussard, P. 'Field techniques for investigations of population regulation in a 'ubiquitous' butterfly'. *J. Lepidopt. Soc.* **25**. 1971, 22–9.
Brussard, P. F. & Ehrlich, P. R. 'The population structure of *Erebia epipsodea* (Lepidoptera: Satyrinae)'. *Ecology* **51**. 1970a, 119–29.
Brussard, P. F. & Ehrlich, P. R. 'Adult behaviour and population structure in *Erebia epipsodea* (Lepidoptera: Satyrinae)'. *Ecology* **51**. 1970b, 880–5.
Bryner, R. 'Dokumentation über den Rückgang der Schmetterlingsfauna in der Region Biel–Seeland–Chasseval'. *Beitr. Natursch. Schweiz.* **9**. 1987, 1–92.
BUTT (Butterflies Under Threat Team). The management of chalk grassland for butterflies. Focus on nature conservation No. 17. Peterborough, Nature Conservancy Council, 1986.
Butterfly Conservation (BC). *Butterflies for the new millenium.* Information Pack. Colchester, Butterfly Conservation, 1996.
Callaghan, C. J. 'A study of isolating mechanisms among neotropical butterflies of the subfamily Riodininae'. *J. Res. Lepid.* **21**. 1983, 159–76.
Calvert, W. H., Zuchowski, W. & Brower, L. P. 'Monarch butterfly conservation: interactions of cold weather, forest thinning and storms on the survival of overwintering monarch butterflies (*Danaus plexippus L.*) in Mexico'. *Atala* **9**. 1984, 2–6.
Cappuccino, N. & Kareiva, P. 'Coping with a capricious environment: a population study of a rare pierid butterfly'. *Ecology* **66**. 1985, 152–61.
Carvalho, J. C. M. & Mielke, O. H. H. 'The trade of butterfly wings in Brazil and its effects upon the survival of the species'. *Proc. XIIIth Int. Congr. Entomol.* **1**. 1971, 486–8.
Chapman, B., Penman, D. & Hicks, P. *Natural Pest Control.* Melbourne, Nelson, 1986.
Chew, S. C. & Robbins, R. K. 'Egg laying in butterflies'. *The Biology of Butterflies.* Vane-Wright, R. I. & Ackery, P. R. (eds). London Academic Press, 1984, 65–79.
Clarke, C. A., Sheppard, P. M. & Scali, V. 'All female broods in the butterfly *Hypolimnas bolina*'. *Proc. R. Soc. Lond. (B)* **189**. 1975, 29–37.
Clench, H. K. 'Behavioural thermoregulation in butterflies'. *Ecology* 47. 1966, 1021–34.
Collins, N. M. *Butterfly houses in Britain. The Conservation Implications.* Gland, IUCN, 1987a.
Collins, N. M. *Legislation to conserve insects in Europe.* Amateur Entomologist's Society Pamphlet No. 13, 1987b.
Collins, N. M. & Morris, M. G. *Threatened Swallowtail Butterflies of the World.* Gland & Cambridge, IUCN, 1985.
Common, I. F. B. & Waterhouse, D. F. *Butterflies of Australia.* Sydney, Angus & Robertson, 1981.
Conservation Committee of Butterfly Conservation (CCBC). 'Lepidoptera restoration: Butterfly Conservation's policy, code of practice and guidelines for action'. *Butterfly Conservation News* **60**. 1995, 20–1.
Cottrell, C. B. 'Aphytophagy in butterflies: its relationship to myrmecophily'. *Zool. J. Linn. Soc.* **80**. 1984, 1–57.
Couchman, L. R. & Couchman, R. 'The butterflies of Tasmania'. *Tasmanian Year Book 1977.* 1977, 66–96.
Courtney, S. P. 'The evolution of egg clustering by butterflies and other insects'. *Amer. Nat.* **123**. 1984, 276–81.
Courtney, S. P. 'The ecology of Pierid butterflies: dynamics and interactions'. *Adv. Ecol. Res.* **15**. 1986, 15–131.
Courtney, S. P. & Chew, F. S. 'Co-existence and host use by a large community of pierid butterflies: habitat is the templet'. *Oecologia* 71. 1987, 210–20.
Craven, I. 'The Arfak Mountains Nature Reserve, Birds Head Region, Irian Jaya, Indonesia'. *Science in New Guinea* **15**. 1989, 47–56.
Crosby, D. F. 'Notes on the Yellowish Skipper *Hesperilla flavescens flavescens* Whs'. *Vic. Ent.* **16**. 1986, 5–7.
Crosby, D. F. 'The Conservation Status of the Eltham Copper Butterfly *Paralucia pyrodiscus lucida* Crosby (Lepidoptera: Lycaenidae)'. Report to Department of Conservation, Forests and Lands, Victoria. 1988.

Crosby, D. F. 'A Management Plan for the Altona Skipper Butterfly *Hesperilla flavescens flavescens* Waterhouse (Lepidoptera: Hesperiidae). Melbourne, *Arthur Rylah Institute for Environmental Research, Tech. Rpt* No. 98, 1990.

Crosby, D. F. & Dunn, K. L. 'The distribution and range extension in Victoria of the butterfly *Ocybadistes walkeri sothis* Waterhouse'. *Vict. Nat.* **106**. 1989, 184–93.

Cushman, J. H. 'The Mission Blue, *Plebejus icarioides missionensis* Hovanitz'. *Conservation biology of Lycaenidae.* New, T. R. (ed.). Gland, IUCN, 1993, 139–40.

Cushman, J. H. & Murphy, D. D. 'Conservation of North American lycaenids — an overview'. *Conservation biology of Lycaenidae.* New, T. R. (ed.). Gland, IUCN, 1993, 37–44.

Daily, G. C. & Ehrlich, P. R. 'Preservation of biodiversity in small rainforest patches: rapid evaluation using butterfly trapping'. *Biodiversity and Conservation* **4**. 1995, 35–55.

Dana, R. P. 'Conservation management of the prairie skippers *Hesperia dacotae*, and *Hesperia ottoe*: basic biology and threat of mortality during prescribed burning in spring'. *Minnesota Agricultural Experiment Stn. Bull.*, No. **594**, 1991.

Davis, B. N. K. 'Pesticides and wildlife conservation'. *J. ent. Soc. Aust. (N.S.W.)* **2**. 1965, 1–7.

Dempster, J. P. 'The natural control of populations of butterflies and moths'. *Biol. Revs* **58**. 1983, 461–81.

Dempster, J. P. 'The ecology and conservation of *Papilio machaon* in Britain'. *Ecology and conservation of butterflies.* Pullin, A. S. (ed.). London, Chapman & Hall, 1995, 137–49.

Dempster, J. P. & Hall, M. L. 'An attempt at re-establishing the swallowtail butterfly at Wicken Fen'. *Ecol. Entomol.* **5**. 1980, 327–34.

Dempster, J. P. & Hall, M. L. 'The swallowtail re-establishment project at Wicken Fen, 1995'. *Butterfly Conservation News* **62**, 1996, 11.

Dennis, R. L. H. *The British butterflies: their origin and establishment.* E. W. Classey, Oxford, 1977.

Dennis, R. L. H. 'Islands, regions, ranges and gradients'. *The ecology of butterflies in Britain'.* Dennis, R. L. H. (ed.) Oxford, Oxford University Press, 1992, 1–21.

Dennis, R. L. H. '*Butterflies and climate change*'. Manchester, University Press, 1993.

Dennis, R. L. H. & Shreeve, T. G. 'Climate change and the British butterfly fauna: opportunities and constraints'. *Biol. Conserv.* **55**. 1991, 1–16.

Dethier, V. G. 'Food plant distribution and density and larval dispersal as factors affecting insect populations'. *Can. Ent.* **88**. 1959, 581–96.

De Vries, P. J. 'The use of epiphylls as larval host plants by the neotropical riodinid butterfly, *Sarota gyas*'. *J. nat. Hist.* **22**. 1988a, 1447–50.

De Vries, P. J. 'Stratification of fruit-feeding nymphalid butterflies in a Costa Rican rainforest'. *J. Res. Lepid.* **26**. 1988b, 98–108.

De Viedma, M. G. & Gomez-Bustillo, M. R. *Libro rojo de los Lepidopteros lbericos.* Madrid, Instituto Nacional para la Conservaci la Naturaleza, 1976.

Dickson, R. *A Lepidopterist's Handbook.* Hanworth, Amateur Entomologist's Society, 1976.

Disney, R. H. L. 'Assessments using invertebrates: posing the problem'. *Wildlife Conservation Evaluation.* Usher, M. B. (ed.). London, Chapman & Hall, 1986, 271–93.

Douwes, P. 'Size of, gain to, and loss from a population of adult *Heodes vigaureae* L'. *Ent. Scand.* **1**. 1970, 263–81.

Douwes, P. 'An area census method for estimating butterfly population numbers'. *J. Res. Lepid.* **15**. 1976, 146–52.

Dover, J. W. 'A method for recording and transcribing observations on butterfly behaviour'. *Ent. Gazette* **40**. 1989, 95–100.

Dover, J. W. 'The conservation of insects on arable farmland'. *The conservation of insects and their habitats.* Collins, N. M. & Thomas, J. A. (eds). London, Academic Press, 1991, 293–317.

Dover, J. W. 'Factors affecting the distribution of satyrid butterflies on arable farmland'. *J. Appl. Ecol.* **33**, 1996, 723–34.

Dover, J., Sotherton, N. & Gobbett, K. 'Reduced pesticide inputs on cereal field margins: the effects on butterfly abundance'. *Ecol. Entomol.* **15**. 1990, 17–24.

Dowdeswell, W. H. *The Life of the meadow brown.* London, Heinemann, 1982.
Downey, J. C. 'Studies on endangered prairie skippers'. *Atala* 7. 1981, 27.
Duffey, E. 'Ecological studies on the Large Copper butterfly *Lycaena dispar* Haw. *batavus* Obth. at Woodwalton Fen National Nature Reserve, Huntingdonshire'. *J. appl. Ecol.* 5. 1968, 69–96.
Duffey, E. 'The re-establishment of the large copper butterfly *Lycaena dispar batava* Obth. at Woodwalton Fen National Nature Reserve, Cambridgeshire, England, 1969–73'. *Biol. Conserv.* **12**. 1977, 143–58.
Dunbar, D. (ed.) *Saving Butterflies. A practical guide to the conservation of butterflies.* Colchester, British Butterfly Conservation Society, 1993.
Ebert, G. & Rennwald, E. *Die Schmetterlinge Baden-Württembergs.* Stuttgart, Ulmer Verlag, 1991.
Edwards, E. D. 'Delayed ovarian development and aestivation in adult females of *Heteronympha merope merope* (Lepidoptera: Satyrinae)'. *J. Aust. ent. Soc.* **12**. 1973, 92–8.
Ehrlich, P. R. 'Intrinsic barriers to dispersal in the checkerspot butterfly, *Euphydryas editha*'. *Science* **134**. 1961, 108–9.
Ehrlich, P. R. & Davidson, S. E. 'Techniques for capture-recapture studies of Lepidoptera populations'. *J. Lepidopt. Soc.* **14**. 1960, 227–9.
Ehrlich, P. R. & Murphy, D. D. 'Conservation lessons from long-term studies of checkerspot butterflies'. *Conserv. Biol.* **1**. 1987, 122–31.
Ehrlich, P. R. & Raven, P. H. 'Butterflies and plants: a study in co-evolution'. *Evolution* **18**. 1965, 586–608.
Ehrlich, P. R., Breedlove, D. E. Brussard, P. F. & Sharp, M. A. 'Weather and the 'regulation' of subalpine populations'. *Ecology* **53**. 1972, 243–7.
Ehrlich, P. R., Murphy, D. D., Singer, M. C. Sherwood, C. B., White, R. R. & Brown, I. L. 'Extinction, reduction, stability and increase: the responses of checkerspot butterfly (*Euphydryas*) populations to the drought'. *Oecologia* **46**. 1980, 101–5.
Ehrlich, P. R., Launer, A. E. & Murphy, D. D. 'Can sex ratio be defined or determined? The case of a population of checkerspot butterflies'. *Amer. Nat.* **124**. 1984, 527–39.
Elmes, G. W. & Thomas, J. A. 'Complexity of species conservation in managed habitats: interaction between Maculinea butterflies and their ant hosts'. *Biodiversity and Conservation* **1**, 1992, 155–69.
Emmel, T. C. & Boender, R. 'Wings in paradise: Florida's Butterfly World'. *Wings* **15**. 1991, 7–12.
Emmel, T. C. & Garraway, E. 'Ecology and Conservation Biology of the Homerus Swallowtail in Jamaica (Lepidoptera: Papilionidae). *Tropical Lepidoptera* **1**. 1990, 63–76.
Emmel, T. C. & Minno, M. C. 'The Atala butterfly, *Eumaeus atala florida* (Röber)'. *Conservation biology of Lycaenidae.* New, T. R. (ed.). Gland, IUCN, 1993, 129–30.
Erhardt, A. 'Diurnal Lepidoptera: sensitive indicators of cultivated and abandoned grassland'. *J. appl. Ecol.* **22**. 1985, 849–62.
Erhardt, A. 'Ecology and conservation of alpine Lepidoptera'. *Ecology and conservation of butterflies.* Pullin, A. S. (ed.). London, Chapman & Hall, 1995, 258–76.
Erhardt, A. & Thomas, J. A. 'Lepidoptera as indicators of change in the seminatural grasslands of lowland and upland Europe'. *The conservation of insects and their habitats.* Collins, N. M. & Thomas, J. A. (eds). London, Academic Press, 1991, 213–36.
Erwin, T. L. 'Tropical forests: their richness in Coleoptera and other arthropod species'. *Coleopt. Bull.* **36**. 1982, 74–5.
Eyre, M. D. & Rushton, S. P. 'Quantification of conservation critera using invertebrates'. *J. appl. Ecol.* **26**. 1989, 159–71.
European Charter for Invertebrates 1986. Strasbourg.
Fast, H. 'On the distribution of Lepidoptera in Great Britain and Ireland'. *Trans. ent. Soc. Lond.* **4**. 1868, 417–517.
Feltham, N. F. The ecology of a Pietermaritzburg butterfly assemblage: a conservation perspective. M.Sc. Thesis, University of Natal, Pietermaritzburg, 1995.
Feltwell, J. *The natural history of butterflies.* London, Croom Helm, 1986.

Feltwell, J. *The conservation of butterflies in Britain, past and present.* Battle, Wildlife Matters, 1995.

Fisher, R. H. *Butterflies of South Australia.* Adelaide, Government Printer, 1978.

Fitter, R. & Fitter, M. (eds). *The Road to Extinction.* Gland & Cambridge, IUCN, 1987.

Fitzpatrick, S. M. & Wellington, W. G. 'Insect Territoriality'. *Can. J. Zool.* **61**. 1983, 471–86.

Ford, E. B. *Butterflies.* London, Collins, 1945.

Fowler, S. V. (+ nine other authors). 'Survey and management proposals for a tropical deciduous forest reserve at Ankarana in northern Madagascar'. *Biol. Conserv.* **47**. 1989, 297–313.

Frazer, J. F. D. 'Estimating butterfly numbers'. *Biol. Conserv.* **5**. 1973, 271–6.

FWAG. *Encouraging butterflies on farms.* Technical Information leaflet. Kenilworth, National Agricultural Centre, 1995.

Gagné, W. C. & Howarth, F. G. 'Conservation status of endemic Hawaiian Lepidoptera'. *Proc. 3rd Congr. eur. Lepid.* 1982, 74–84.

Gall, L. F. 'Population structure and recommendations for conservation of the narrowly endemic alpine butterfly, *Boloria acrocnema* (Lepidoptera: Nymphalidae)'. *Biol. Conserv.* **28**. 1984a, 111–38.

Gall, L. F. 'The effects of capturing and marking on subsequent activity in *Boloria acrocnema* (Lepidoptera: Nymphalidae) with a comparison of different numerical models that estimate population size'. *Biol. Conserv.* **28**. 1984b, 139–54.

Gall, L. F. 'Measuring the size of Lepidopteran populations'. *J. Res. Lepid.* **24**. 1985, 97–116.

Gall, L. F. & Sperling, F. A. H. 'A new high altitude species of *Boloria* from southwestern Colorado (Nymphalidae), with a discussion of phenetics and hierarchical considerations'. *J. Lepidopt. Soc.* **34**. 1980, 230–52.

Gardiner, B. O. C. '*Pieris brassicae* L. established in Chile; another Palaearctic pest crosses the Atlantic (Pieridae)'. *J. Lepidopt. Soc.* **28**. 1973, 269–77.

Gennardus, Br. 'Enkele waarnemingen aan verschillende Kweken van rupsen en vlinders na een vulkanische asregen op Java (Lepidoptera)'. *Ent. Ber.* **43**. 1983, 69–71.

Gibbs, G. W. *New Zealand Butterflies.* Auckland, Collins, 1980.

Gilbert, L. E. 'Ecological consequences of a coevolved mutualism between butterflies and plants'. *Coevolution of Animals and Plants.* Gilbert, L. E. & Raven, P. H. (eds). Austin, University of Texas Press, 1975, 210–40.

Gilbert, L. E. 'The biology of butterfly communities'. *The Biology of Butterflies.* Vane-Wright, R. I. & Ackery, P. R. (eds). London, Academic Press, 1984, 41–54.

Gilbert, L. E. & Singer, M. C. 'Butterfly Ecology'. *Annu. Rev. Ecol. Syst.* **6**. 1975, 365–97.

Goss, H. 'On the probable early extinction of *Lycaena arion* in Britain'. *Entomol. mon. Mag.* **21**. 1884, 107–9.

Hama, E., Ishii, M. & Sibatani, A. (eds). *Decline and Conservation of Butterflies in Japan.* I. Osaka Lepidopterological Society of Japan, 1989.

Hammond, P. C. & McCorkle, D. V. 'The decline and extinction of *Speyeria* populations resulting from human environmental disturbances (Nymphalidae: Argynninae). *J. Res. Lepid.* **22**. 1984, 217–24.

Hanski, I. & Gilpin, M. 'Metapopulation dynamics: brief history and conceptual domain'. *Biol. J. Linn. Soc.* **42**. 1991, 3–16.

Harding, P. T., Asher, J. & Yates, T. J. 'Butterfly monitoring I — recording the changes'. *Ecology and conservation of butterflies.* Pullin, A. S. (ed.). London, Chapman & Hall, 1996, 3–22.

Hardy, P. B. 'Monitoring of selected butterfly species and their host plant-habitat by 100 m^2 units'. *Ent. Gaz.* **45**. 1994, 159–64.

Harrison, S. 'Long-distance dispersal and colonization in the bay checkerspot butterfly, *Euphydryas editha bayensis*'. *Ecology* **70**. 1989, 1236–43.

Harrison, S., Murphy, D. D. & Ehrlich, P. R. 'Distribution of the Bay Checkerspot butterfly, *Euphydryas editha bayensis*: evidence for a metapopulation model'. *Amer. Nat.* **132**. 1988, 360–82.

Hastings, A. & Harrison, S. 'Metapopulation dynamics and genetics'. *Ann. Rev. Ecol. Syst.* **25**. 1994, 167–88.

Heath, J. *Provisional atlas of the insects of the British Isles.* Part 1. Lepidoptera: Rhopalocera (Butterflies). (Maps 1–57). Huntingdon Nature Conservancy, 1970.

Heath, J. 'A century of change in the Lepidoptera'. *The changing flora and fauna of Britain.* Hawksworth, D. L. (ed.) London, Academic Press, 1974, 275–92.

Heath, J. *Threatened Rhopalocera (Butterflies) of Europe.* Council of Europe, Strasbourg, 1981.

Heath, J. Pollard, E. & Thomas, J. A. *An atlas of the butterflies of the British Isles.* London, Penguin Press, 1984.

Henning, G. A. & Henning, S. F. 'Conservation of Lepidoptera in southern Africa'. *A practical guide to butterflies and moths in southern Africa.* Woodhall, S. E. (ed.). Florida Hills, The Lepidopterists' Society of southern Africa. 1992, 29–42.

Henning, S. F. 'Biological groups within the Lycaenidae'. *J. ent. Soc. sth Afr.* **46**. 1983, 65–85.

Henning, S. F. & Henning, G. A. 'South African Red Data Book — Butterflies'. Report No. 158. South African National Scientific Programmes, 1989.

Hilburn, D. J. 'A non-migratory, non-diapausing population of the Monarch butterfly, *Danaus plexippus* (Lepidoptera: Danaidae) in Bermuda'. *Fla Entomol.* **72**. 1989, 494–9.

Hill, P. & Twist, C. *Dragonflies and butterflies. A site guide.* Chelmsford, Arlequin, 1996.

Hirukawa, N. 'Degradation of *Sasakia charonda* populations caused by environmental alterations of its habitat in Matsushiro town, Nagano Prefecture'. *Decline and conservation of Butterflies in Japan,* **II**. 1993, 129–32 (in Japanese).

Holl, K. D. 'The effect of coal surface mine reclamation on diurnal lepidopteran conservation'. *J. Appl. Ecol.* **33**. 1996, 225–36.

Holland, W. J. *The Butterfly Book.* New York, Doubleday, 1898.

Holloway, J. D. *Moths of Borneo with special reference to Mt. Kinabalu.* Kuala Lumpur, Malayan Nature Society, 1976.

Hooper, M. D. 'The size and surroundings of nature reserves'. *The scientific management of animal and plant communities for conservation.* Duffey, E. & Watt, A. S. (eds). Oxford, Blackwell, 1971, 555–61.

Hoskins, A. & Hardy, P. 'The butterfly trade in Malaysia'. *Butterfly Conservation News* **55**. 1993, 28–31.

Ilse, D. 'New observations on responses to colours in egg-laying butterflies'. *Nature* **140**. 1937, 544–5.

Ishii, M. 'Decline and conservation of butterflies in Japan'. *Decline and conservation of butterflies in Japan,* **III**. 1996, 157–67.

IUCN *Red List of Threatened Animals.* Gland & Cambridge, IUCN, 1988, 1994.

IUCN. *IUCN Red List Categories.* Gland, IUCN. 1994.

Jackson, B. S. 'Thoughts on providing nectar sources for some North American butterflies'. *Atala* **13**. 1986, 8–13.

JCCBI 1971. A code for insect collecting [published in several entomological journals and as separate leaflet: e.g. *Entomol. mon. Mag.* **107**: 193–5].

JCCBI. Insect re-establishment: a code of conservation practices. *Antenna* **10**. 1986, 13–18.

Kendall, R. O. 'Larval food plants and life history notes for eight months from Texas and Mexico'. *J. Lepidopt. Soc.* **30**. 1976, 264–71.

Khoo, S. N. & Chng, W. W. *Penang Butterfly Farm.* Penang, Yeoh Teow Giap, 1987.

Kingsolver, J. G. 'Thermal ecology of *Pieris* butterflies (Lepidoptera: Pieridae): a new mechanism of behavioural thermoregulation'. *Oecologia* **66**. 1985a, 540–5.

Kingsolver, J. G. 'Thermoregulatory significance of wing melanization in *Pieris* butterflies (Lepidoptera: Pieridae): physics, posture, and pattern'. *Oecologia* **66**. 1985b, 546–53.

Knapton, R. W. 'Lek structure and territoriality in the chryxus arctic butterfly, *Oeneis chryxus* (Satyridae)'. *Behav. Ecol. Sociobiol.* **17**. 1985, 389–95.

Kremen, C. 'Assessing the indicator properties of species assemblages for natural areas monitoring'. *Ecol. Applns* **2**. 1992, 203–17.

Kremen, C. 'Biological inventory using target taxa: a case study of the butterfies of Madagascar'. *Ecol. Applns* **4**. 1994, 407–22.

Kudrna, O. *Butterflies of Europe. 8. Aspects of the Conservation of Butterflies in Europe.* Weisbaden, Aula-Verlag, 1986.

Kudrna, O. 'Conservation of butterflies in central Europe'. *Ecology and conservation of butterflies.* Pullin, A. S. (ed.). London, Chapman & Hall, 1995, 248–57.

Kudrna, O., Lukasek, P. and Slavik, K. 'Zur ergfolgreichen Wiederansiedlung von *Parnassius apollo* (Linnaeus 1758) in Tschechien'. *Oedippus* **9**. 1994. 1–37.

Lane, J. 'The status of Monarch butterfly overwintering sites in Alta California'. *Atala* **9**. 1984, 17–20.

Langston, R. L. 'Extended flight periods of coastal and dune butterflies in California'. *J. Lepidopt. Soc.* **13**. 1974, 83–98.

Larsen, T. B. 'Butterfly art in Africa — conservation implications'. *Butterfly Conservation News* No. **61**. 1995a, 12–14.

Larsen, T. B. 'Butterfly biodiversity and conservation in the Afrotropical region'. *Ecology and conservation of butterflies.* Pullin, A. S. (ed.). London, Chapman & Hall, 1995b, 290–303.

Larsen, T. B. 'Hazards of butterfly collecting — 'from rat trap to barbeque bottom', Jamaica, February 1994'. *Ent. Rec. J. Var.* **107**. 1995c, 150–2.

Larsen, T. B., Riley, J. & Cornes, M. A. 'The butterfly fauna of a secondary bush locality in Nigeria'. *J. Res. Lepid.* **18**. 1979, 4–23.

Latheef, M. A. & Ortiz, J. H. 'Influence of companion plants on oviposition of imported cabbageworm, *Pieris rapae* (Lepidoptera: Pieridae), and cabbage looper, *Trichoplusia ni* (Lepidoptera: Noctuidae), on collard plants'. *Canad. Ent.* **115**. 1983, 1529–31.

Launer, A. E. & Murphy, D. D. 'Umbrella species and the conservation of habitat fragments: a case of a threatened butterfly and a vanishing grassland ecosystem'. *Biol. Conserv.* **69**. 1994, 145–53.

Lepidopterists' Society. Statement on collecting policy, 1982.

Levins, R. A. 'Extinction'. *Lectures on mathematics in the life sciences. 2.* Rhode Island, Providence, American Mathematical Society, 1970. 77–107.

Ligue Suisse pour la Protection de la Nature. *Les papillons de jour et leurs biotopes. Especes. Dangers qui les menacent. Protection.* Basle, 1987.

Lockwood, J. A. 'The moral standing of insects and the ethics of extinction'. *fla Entomol.* **70**. 1987, 70–89.

Mackay, D. A. & Jones, R. E. 'Leaf shape and the host-finding behaviour of two ovipositing monophagous butterfly species'. *Ecol. Entomol.* **14**. 1989, 423–31.

Main, A. R. 'Rare species: precious or dross?' *Species at risk: research in Australia.* Groves, R. H. & Ride, W. D. L. (eds). Canberra, Australian Academy of Science, 1982, 163–74.

Makibayashi, I. 'Recent progress on the conservation effort for *Sasakia charonda* (the Great Purple Emperor) in Ranzanmachi, Saitama Prefecture'. *Decline and conservation of butterflies in Japan.* **III**. 1996, 176–9.

Malcolm, S. B. 'Conservation of Monarch butterfly migration in North America: an endangered phenomenon'. *Biology and conservation of the Monarch butterfly.* Malcolm, S. B. & Zalucki, M. P. (eds). Los Angeles, Natural History Museum of Los Angeles County, 1996, 357–61.

Malcolm, S. B. & Zalucki, M. P. (eds). *Biology and conservation of the Monarch butterfly.* Los Angeles, Natural History Museum of Los Angeles County, 1996.

Marshall, A. G. 'The butterfly industry of Taiwan'. *Antenna.* **6**. 1982, 203–4.

Mattoni, R. 'Conflict and conservation: the El Segundo Blue and the airport (Lycaenidae)'. *Nota lepid.* **12**, Supplement 1. 1989, 12.

Mattoni, R. H. T. 'The endangered El Segundo blue butterfly'. *J. Res. Lepid.* **29**. 1992, 277–304.

Mattoni, R. H. T. 'The Palos Verdes blue, *Glaucopsyche lygdamus palosverdesensis* Perkins and Emmel'. *Conservation biology of Lycaenidae.* New, T. R. (ed.). Gland, IUCN, 1993, 135–6.

McCabe, T. L. 'The Dakota Skipper, *Hesperia dacotae* (Skinner): range and biology with special reference to North Dakota'. *J. Lepidopt. Soc.* **35**. 1981, 179–93.

Mercer, C. W. L. 'Survey of Queen Alexandra's birdwing butterfly on the Managalase Plateau of Oro Province, Papua New Guinea — 18th–28th June 1982'. Lae, Papua New Guinea University of Technology, Dept. of Forestry. 1992.

Miskin, W. H. 'Occurrence of *Danaus plexippus* in Queensland'. *Entomol. mon. Mag.* **8**. 1871, 17.

Mitchell, G. A. *The national butterflies of Papua New Guinea*. Papua New Guinea Department of Natural Resources, Wildlife Branch. n.d.

Morris, M. G. 'Conservation and the collector'. *The Moths and Butterflies of Great Britain and Ireland*. I. Heath, J. (ed.) Oxford and London, Blackwell and Curwen, 1976.

Morris, M. G. 'The scientific basis of insect conservation'. *Proc. 3rd European Congr. Entomol. (Amsterdam)* **3**. 1986, 357–67.

Morris, M. G. 'Changing attitudes to nature conservation: the entomological perspective'. *Biol. J. Linn. Soc.* **32**. 1987, 213–23.

Morris, M. G. & Webb, N. R. 'The importance of field margins for the conservation of insects'. *BCPC Monograph* No. 35. 1987, 53–65.

Morton, A. C. 'Rearing butterflies on artifical diets'. *J. Res. Lepid.* **18**. 1981, 221–7.

Morton, A. C. 'The effects of marking and capture on recapture frequencies of butterflies'. *Oecologia* **53**. 1982, 105–10.

Morton, A. C. 'Butterfly conservation — the need for a captive breeding institute'. *Biol. Conserv.* **25**. 1983, 19–33.

Morton, A. C. 'The effects of marking and handling on recapture frequencies of butterflies'. *The Biology of Butterflies*. Vane-Wright, R. I. & Ackery, P. R. (eds). London, Academic Press, 1984.

Mouffet, T. *The Theater of Insects*. London, 1658.

Mueller, L. E., Wilcox, B. A., Ehrlich, P. R. & Murphy, D. D. 'A direct assessment of the role of genetic drift in determining allele frequency variation in populations of *Euphydryas editha*'. *Genetics* **110**. 1985, 495–511.

Muggleton, J. & Benham, B. R. 'Isolation and the decline of the Large Blue butterfly (*Maculinea arion*) in Great Britain'. *Biol. Conserv.* 7. 1975, 119–28.

Munguira, M. L. 'Conservation of butterfly habitats and diversity in European Mediterranean countries'. *Ecology and conservation of butterflies*. Pullin, A. S. (ed.). London, Chapman & Hall, 1995, 277–89.

Munguira, M. L. & Martin, J. 'The conservation of endangered lycaenid butterflies in Spain'. *Biol. Conserv.* **66**. 1993, 17–22.

Munguira, M. L. & Thomas, J. A. 'Use of road verges by butterfly and burnet populations, and the effect of roads on adult dispersal and mortality'. *J. Appl. Ecol.* **29**. 1992, 316–29.

Munguira, M. L., Martin, J. & Balletto, E. 'Conservation biology of Lycaenidae: a European overview'. *Conservation biology of Lycaenidae*. New, T. R. (ed.). Gland, IUCN, 1993, 23–34.

Murata, K. & Nohara, K. 'Decline and conservation of *Shijimiaeoides divinus* (Matsumara) in Kumamoto Prefecture'. *Decline and conservation of butterflies in Japan*, **II**. 1993, 151–9 (in Japanese).

Murphy, D. D. 'Are we studying our endangered butterflies to death?' *J. Res. Lepid.* **26**. 1989, 236–9.

Murphy, D. D. & Weiss, S. B. 'Ecological studies and the conservation of the Bay Checkerspot butterfly, *Euphydryas editha bayensis*'. *Biol. Conserv.* **46**. 1988, 183–200.

Murphy, D. D. & Weiss, S. B. 'A long-term monitoring plan for a threatened butterfly'. *Conserv. Biol.* **2**. 1988, 367–74.

Murphy, D. D. & Wilcox, B. A. 'Butterfly diversity in natural habitat fragments: a test of the validity of vertebrate-based management'. *Wildlife 2000. Modelling habitat relationships of terrestrial vertebrates*. Verner, J., Morrison, M. L. & Ralph, C. J. (eds). Madison, University of Wisconsin Press, 1986, 287–92.

Murphy, D. D., Freas, K. E. & Weiss, S. B 'An environment-metapopulation approach to population viability analysis for a threatened invertebrate'. *Conserv. Biol.* 4. 1990, 41–51.

Nagano, C. D. & Sakai, W. H. 'Making the world safe for monarchs'. *Outdoor California* **49**. 1988, 5–9.

Nakamura, I. & Ae, S. A. 'Prolonged pupal diapause of *Papilio alexanor*: arid zone adaptation directed by larval host plant'. *Ann. ent. Soc. America* **70**. 1977, 481–4.

National Research Council. *Butterfly farming in Papua New Guinea*. Washington, D.C. National Academy Press, 1983.

Nelson, S. M. & Andersen, D. C. 'An assessment of riparian environmental quality by using butterflies and disturbance susceptibility scores'. *Southw. Nat.* **39**. 1994, 137–42.

New, T. R. *Insect Conservation: an Australian perspective*. Dordrecht, W. Junk, 1984.

New, T. R. *Butterfly Conservation*. Melbourne, Entomological Society of Victoria, 1987.

New, T. R. (ed.) *Conservation biology of Lycaenidae*. Gland, IUCN. 1993.

New, T. R. 'Butterfly ranching: sustainable use of insects and sustainable benefits to habitats'. *Oryx* **28**. 1994, 169–72.

New, T. R. *Introduction to invertebrate conservation biology*. Oxford, Oxford University Press. 1995.

New, T. R. 'Evaluating the status of butterflies for conservation'. *Decline and Conservation of Butterflies in Japan* **III**. 1996, 4–21.

New, T. R. & Britton, D. R. 'Refining a conservation plan for an endangered lycaenid butterfly, *Acrodipsas myrmecophila*, in Victoria, Australia'. *J. Ins. Conserv.* (in press).

New, T. R. & Collins, N. M. *Swallowtail butterflies: an action plan for their conservation*. IUCN, Gland and Cambridge, 1991.

New, T. R. & Thornton, I. W. B. 'The butterflies of Anak Krakatau, Indonesia: faunal development in early succession'. *J. Lepidopt. Soc.* **46**. 1992, 83–96.

New, T. R. & Yen, A. L. 'Species management and recovery plans for butterflies (Insecta: Lepidoptera) in Australia'. *People and nature conservation. Perspectives on private land use and endangered species recovery*. Bennett, A., Backhouse, G. & Clark, T. (eds). Mosman, Royal Zoological Society of New South Wales, 1995, 15–21.

New, T. R., Bush, M. B., Thornton, I. W. B. & Sudarman, H. K. 'The butterfly fauna of the Krakatau Islands after a century of colonization'. *Phil. Trans. R. Soc. Lond. B* **322**. 1988, 445–57.

New, T. R., Britton, D. R., Hinkley, S. D. & Miller, L. J. 'The ant fauna of Mount Piper and its relevance to environmental assessment and the conservation of a threatened invertebrate community'. Flora and Fauna Technical Report No. 143, Department of Natural Resources and Environment, Victoria, 1996.

New, T. R., Pyle, R. M., Thomas, J. A., Thomas, C. D. & Hammond, P. C. 'Butterfly Conservation Management'. *Annu. Rev. Entomol.* **40**. 1995, 57–83.

Newman, L. H. *Create a Butterfly Garden*. London, John Baker, 1967.

Neyland, M. '*The Ptunarra brown butterfly Oreixenica ptunarra. Conservation Research Statement*'. Hobart, Tasmanian Department of Parks, Wildlife and Heritage, 1992.

Neyland, M. 'The ecology and conservation management of the Ptunarra brown butterfly *Oreixenica ptunarra* (Lepidoptera; Nymphalidae; Satyrinae) in Tasmania, Australia'. *Pap. Proc. Roy. Soc. Tasm.* **127**, 1993, 43–8.

Nielsen, E. T. 'On the habits of the migratory butterfly, *Ascia monuste*'. *Biol. Meddr.* **23**. 1961, 1–81.

Nielsen, V. & Monge-Nájera, J. 'A comparison of four methods to evaluate butterfly abundance, using a tropical community'. *J. Lepidopt. Soc.* **45**. 1991, 241–3.

Nikusch, I. 'First trials to save threatened populations of *Parnassius apollo* by transplantation to new suitable biotopes'. *Proc. Third European Congress of Lepidopterology*. 1982.

Oates, M. R. 'Butterfly conservation within the management of grassland habitats'. *Ecology and conservation of butterflies*. Pullin, A. S. (ed.). London, Chapman & Hall, 1995, 98–112.

Olliff, A. S. *Australian Butterflies: a brief account of the native families*. Sydney, Natural History Association of New South Wales, 1889.

Opler, P. A. 'North American problems and perspectives in insect conservation'. *The conservation of insects and their habitats.* Collins, N. M. & Thomas, J. A. (eds). London, Academic Press. 1991. 9–32.

Opler, P. A. 'Conservation and management of butterfly diversity in North America'. *Ecological conservation of butterflies.* Pullin, A. S. (ed). London, Chapman & Hall. 1995. 316–24.

Orsak, L. 'San Bruno's $800 million butterflies: the Habitat Conservation Plan'. *Wings* **9**. 1982a, 1–16.

Orsak, L. J. 'The endangered Mission Blue butterfly of California — indicator of an imperiled natural ecosystem'. Xerces Society Educational leaflet, No. 8. 1982b.

Orsak, L. 'The endangered El Segundo Blue butterfly of California and its dune habitat'. Xerces Society Educational Leaflet, No. 9. 1982c.

Orsak, L. 'Killing butterflies to save butterflies: a tool for tropical forest conservation in Papua New Guinea'. *Newsl. Lepidopt. Soc.* No. **3**. 1993, 71–80.

Otero, L. S. & Brown, K. S., Jr. 'Biology and ecology of *Parides ascanius* (Cramer 1775) (Lep., Papilionidae), a primitive butterfly threatened with extinction'. *Atala* **10–12**. 1986, 2–16.

Owen, D. F. 'A further analysis of the insect records from the London bombed sites'. *Ent. Gazette* 5. 1954, 51–60.

Owen, D. F. 'Inheritance of sex-ratio in the butterfly *Acraea encedon*'. *Nature* **225**. 1970, 662–3.

Owen, D. F. *Tropical Butterflies*. Oxford, Clarendon Press, 1971a.

Owen, D. F. 'Species diversity in butterflies in a tropical garden'. *Biol. Conserv.* **3**. 1971b, 191–8.

Owen, D. F. 'Estimating the abundance and diversity of butterflies'. *Biol. Conserv.* **8**. 1975, 173–83.

Owen, D. F. 'Insect diversity in an English suburban garden'. *Perspectives in Urban Entomology*. Frankie, G. W. & Koehler, C. S. (eds). New York, Academic Press, 1978, 13–29.

Packer, L. 'The status of two butterflies, Karner Blue (*Lycaeides melissa samuelis*) and Frosted Elfin (*Incisalia irus*), restricted to oak savannah in Ontario'. *Conserving Carolinian Canada* Allen, G. W., Eagles, P. F. J. & Price, S. W. (eds). Waterloo, University of Waterloo Press, 1991.

Parsons, M. J. 'Farming manual: insect farming and trading agency'. Papua New Guinea. Division of Wildlife, 1978.

Parsons, M. J. 'The biology and conservation of *Ornithoptera alexandrae*'. *The Biology of Butterflies*. Vane-Wright, R. I. & Ackery, P. R. (eds). London, Academic Press, 1984, 327–31.

Parsons, M. J. *Butterflies of the Bulolo-Wau Valley*. Honolulu, Bishop Museum Press, 1992a.

Parsons, M. J. The butterfly farming and trading industry in the Indo-Australian region and its role in tropical forest conservation. *Trop. Lepidoptera* 3 Supplement 1, 1992b, 1–31.

Parsons, M. J. 'The world's largest butterfly endangered, the ecology, status and conservation of *Ornithoptera alexandrae* (Lepidoptera: Papilionidae)'. *Trop. Lepidoptera* 3 Supplement 1, 1992c. 33–60.

Parsons, M. J. Butterfly farming in the Indo-Australian region: an effective and sustainable means of combining conservation and commerce to protect tropical forests. *Decline and conservation of butterflies in Japan* III. 1996a, 63–77.

Parsons, M. J. Conservation of the birdwing butterflies (*Ornithoptera* and *Troides*, Lepidoptera: Papilionidae): not hard if we try. *Decline and Conservation of Butterflies in Japan* III. 1996b, 150–6.

Pearman, G. I. (ed.) *Greenhouse. Planning for climatic change*. Melbourne, CSIRO, 1989.

Pearse, F. K. & Murray, N. D. 'Clinal variation in the Common Brown butterfly *Heteronympha merope merope* (Lepidoptera: Satyrinae)'. *Aust. J. Zool.* **29**. 1981, 631–47.

Pierce, N. E. 'Amplified species diversity: a case study of an Australian lycaenid butterfly and its attendant ants'. *The biology of butterflies*. Vane-Wright, R. I. & Ackery, P. R. (eds). London, Academic Press, 1984, 197–200.

Pinhey, E. *Butterflies of southern Africa.* Cape Town, Nelson, 1965.
Pljushtch, I. G. 'Notes on a little known ecologically displaced blue, *Agriades pyrenaicus ergane* Higgins (Lycaenidae)'. *J. Res. Lepid.* 27 (1988) 129–34.
Pollard, E. 'A method for assessing changes in the abundance of butterflies'. *Biol. Conserv.* **12**. 1977, 115–34.
Pollard, E. 'Population ecology and change in range of the white admiral butterfly *Ladoga camilla* L. in England'. *Ecol. Entomol.* **4**. 1979, 61–74.
Pollard, E. 'Monitoring butterfly abundance in relation to the management of a nature reserve'. *Biol. Conserv.* **24**. 1982, 317–28.
Pollard, E. 'Temperature, rainfall and butterfly numbers'. *J. appl. Ecol.* **25**. 1988, 819–28.
Pollard, E. & Yates, T. J. *Monitoring butterflies for ecology and conservation.* London, Chapman and Hall. 1993.
Pollard, E., Elias, D. O., Skelton, M. J. & Thomas, J. A. 'A method of assessing the abundance of butterflies in Monks Wood National Nature Reserve in 1973'. *Ent. Gazette* **26**. 1975, 79–88.
Powell, J. A. & Parker, M. W. 'Lange's Metalmark, *Apodemia mormo langei* Comstock'. *Conservation biology of Lycaenidae.* New, T. R. (ed.). Gland, IUCN. 1993, 116–9.
Pratt, C. 'A modern review of the demise of *Aporia crataegi* L.: the Black-Veined White'. *Ent. Rec. J. Var.* **95**. 1983, 45–52, 161–6, 232–7.
Prendergast, J.R. & Eversham, B. C. 'Butterfly diversity in southern Britain: hotspot losses since 1930'. *Biol. Conserv.* **72**. 1995, 109–14.
Preston-Mafham, R. & Preston-Mafham, K. *Butterflies of the world.* London, Blandford Press, 1988.
Prince, G. B. *The habitat requirements and conservation status of Tasmanian endemic butterflies.* Hobart, Department of Lands, Parks and Wildlife, 1988.
Pullin, A. S., McLean, I. F. G. & Webb, M. R. 'Ecology and conservation of *Lycaena dispar*: British and European perspectives'. *Ecology and conservation of butterflies.* Pullin, A. S. (ed.). London, Chapman & Hall, 1995, 150–64.
Pyle, R. M. 'Conservation of Lepidoptera in the United States'. *Biol. Conserv.* **9**. 1976, 55–75.
Pyle, R. M. 'Butterfly ecogeography and biological conservation in Washington'. *Atala* **8**. 1982, 1–26.
Pyle, R. M. *Handbook for butterfly watchers.* Boston, Houghton Mifflin, 1992.
Pyle, R. M. 'A history of Lepidoptera conservation, with special reference to its Remingtonian debt'. *J. Lepidopt. Soc.* **49**. 1995, 397–411.
Pyle, R. M., Bentzien, M. & Opler, P. A. 'Insect Conservation'. *Annu. Rev. Entomol.* **26**. 1981, 233–58.
Pyle, R. M. & Hughes, S. A. 'Conservation and utilisation of the insect resources of Papua New Guinea'. Papua New Guinea, Department of Natural Resources, 1978.
Rainbow, W. J. *A guide to the study of Australian butterflies.* Melbourne, Lothian, 1907.
Rands, M. R. W. & Sotherton, N. W. 'Pesticide use on cereal crops and changes in the abundance of butterflies on arable farmland in England'. *Biol. Conserv.* **36**. 1986, 71–82.
Ravenscroft, N. O. M. & Warren, M. S. (comp.) Species Action Plan. Northern Brown Argus *Aricia artaxerces.* Wareham, Butterfly Conservation, 1996.
Riotte, J. C. & Uchida, G. 'Butterflies of the Hawaiian Islands according to the stand of late 1976'. *J. Res. Lepid.* 17. 1978, 33–9.
Robbins, R. K. 'How many butterfly species?' *News Lepidopt. Soc.* 1982, 40–1.
Robertson, P. A., Woodburn, M. I. A. & Hill, D. A. 'The effects of woodland management for pheasants on the abundance of butterflies in Dorset, England'. *Biol. Conserv.* **45**. 1988, 159–67.
Ross, G. N. 'Butterfly Magic'. *Wildlife Conservation* **99**. 1996, 20–7.
Rothschild, M. & Farrell, C. *The butterfly gardener.* London, Michael Joseph/Rainbird, 1983.
Ruszczyk, A. 'Mortality of *Papilio scamander scamander* (Lepidoptera: Papilionidae) pupae in four districts of Porto Alegre (S. Brazil) and the causes of superabundance of some butterflies in urban areas. *Revta bras. Biol.* **46**. 1986, 567–79.

Ruszczyk, A. 'Distribution and abundance of butterflies in the urbanization zones of Porto Alegre, Brazil'. *J. Res. Lepid.* **25** (1986). 1987, 157–78.

Ruszczyk, A. & Araujo, A. M. 'Gradients in butterfly species diversity in an urban area in Brazil'. *J. Lepidopt. Soc.* **46**. 1992, 255–64.

Samways, M. J. 'Threatened Lycaenidae of South Africa'. *Conservation biology of Lycaenidae.* New, T. R. (ed.) Gland, IUCN, 1993, 62–9.

Samways, M. J. *Insect conservation biology.* London, Chapman and Hall, 1994.

Schiotz, A. 'The Biology of Extinction'. *The Road to Extinction.* Fitter, R. & Fitter, M. (eds). Gland & Cambridge, IUCN, 1987, 68–70.

Scoble, M. J. 'The structure and affinities of the Hedyloidea: a new concept of the butterflies'. *Bull. Br. Mus. nat. Hist (Ent.)* **53**. 1986, 251–86.

Scott, J. A. 'Mate-locating behaviour in butterflies'. *Amer. Midl. Nat.* **91**. 1974, 103–17.

Scott, J. A. 'Flight patterns among eleven species of diurnal Lepidoptera'. *Ecology* **56**. 1975, 1367–77.

Scriber, J. M. 'Latitudinal gradients in larval feeding specialization of the world Papilionidae (Lepidoptera)'. *Psyche. Camb. Mass.* **80**. 1973, 355–73.

Scudder, S. H. *Butterflies, their structure, changes and life-histories with special references to American forms.* New York, Holt, 1889a.

Scudder, S. H. 'The fossil butterflies of Florissant'. *Ann. Rept U.S. Geol. Survey* pt. 1. 1889b, 433–74.

Shapiro, A. M. 'The temporal component of butterfly species diversity'. *Ecology and evolution of communities.* Cody, M. L. & Diamond, J. M. (eds). Cambridge, Harvard University Press, 1975, 181–95.

Shapiro, A. M. 'Seasonal polyphenisms'. *Evol. Biol.* **9**. 1976, 259–333.

Shapiro, A. M. & Shapiro, A. R. 'The ecological associations of the butterflies of Staten Island'. *J. Res. Lepid.* **12**. 1973, 65–128.

Shapiro, A. 'Extirpation and recolonization of the Buckeye, *Junonia coenia* (Nymphalidae) following the northern California freeze of December, 1990'. *J. Res. Lepid.* **30**. 1991, 209–20.

Sheldon, W. G. 'The destruction of British butterflies'. *Entomologist* **58**. 1925, 105–12.

Shields, O. 'Hilltopping'. *J. Res. Lepid.* **6**. 1967, 69–178.

Shields, O. 'Fossil butterflies and the evolution of Lepidoptera'. *J. Res. Lepid.* **15**. 1976, 132–43.

Shields, O. 'World numbers of butterflies'. *J. Lepidopt. Soc.* **43**. 1989, 178–83.

Shreeve, T. G. 'Habitat selection, mate location, and microclimatic constraints on the activity of the speckled wood butterfly *Pararge aegeria*'. *Oikos* **42**. 1984, 371–7.

Shreeve, T. G. 'Monitoring butterfly movements'. *The ecology of butterflies in Britain.* Dennis, R. L. H. (ed.). Oxford, Oxford University Press, 1992, 120–38.

Shreeve, T. G. & Mason, C. F. 'The number of butterfly species in woodlands'. *Oecologia* **45**. 1980, 414–18.

Shuey, J. A. 'Habitat associations of wetland butterflies near the glacial maxima in Ohio, Indiana and Michigan'. *J. Res. Lepid.* **24**. 1985, 176–86.

Shuey, J. A. & Peacock, J. W. 'Host Plant Exploitation by an Oligophagous Population of *Pieris virginiensis* (Lepidoptera: Pieridae)'. *Amer. Midl. Nat.* **122**. 1989, 255–61.

Sibatani, A. 'Decline and conservation of butterflies in Japan'. *Decline and Conservation of Butterflies in Japan. 1.* Hama, E., Ishii, M. & Sibatani, A. (eds). Osaka, Lepidopterological Society of Japan, 1989, 16–22.

Sikes, D. S. & Ivie, M. A. 'Predation of *Anetia briarea* Godart (Nymphalidae: Danainae) at aggregation sites: a potential threat to the survival of a rare montane butterfly in the Dominican Republic'. *J. Lepidopt. Soc.* **49**. 1995, 223–33.

Silberglied, R. L. 'Visual communication and sexual selection among butterflies'. *The Biology of Butterflies.* Vane-Wright, R. I. & Ackery, P. R. (eds). London, Academic Press, 1984, 207–24.

Sims, S. R. 'Diapause dynamics and host plant suitability of *Papilio zelicaon*'. *Amer. Midl. Nat.* **103**. 1980, 375–84.

Singer, M. C. 'Butterfly-hostplant relationships: host quality, adult choice and larval success'. *The Biology of Butterflies*. Vane-Wright, R. I. & Ackery, P. R. (eds). London, Academic Press, 1984, 81–8.

Singer, M. C. & Gilbert, L. E. 'Ecology of butterflies in the urbs and suburbs'. *Perspectives in Urban Entomology*. Frankie, G. W. & Koehler, C. S. (eds). New York, Academic Press, 1978, 1–11.

Singer, M. C. & Wedlake, P. 'Capture does affect probability of recapture in a butterfly species'. *Ecol. Entomol.* **6**. 1981, 215–16.

Singer, M. C., Ehrlich, P. R. & Gilbert, L. E. 'Butterfly feeding on a lycopsid. *Science*' **172**. 1971, 1341–2.

Smithers, C. N. 'Seasonal distribution and breeding status of *Danaus plexippus* (L.) (Lepidoptera: Nymphalidae) in Australia'. *J. Aust. ent. Soc.* **16**. 1977, 175–84.

Snook, L. C. 'Conservation of the Monarch butterfly reserves in Mexico: focus on the forest'. *Biology and conservation of the Monarch butterfly*. Malcolm, S. B. & Zalucki, M. P. (eds). Los Angeles, Natural History Museum of Los Angeles County, 1996, 363–75.

Southwood, T. R. E. 'Habitat, the templet for ecological strategies?' *J. Anim. Ecol.* **46**. 1977, 337–66.

Southwood, T. R. E. *Ecological methods, with particular reference to the study of insect populations*. London, Chapman & Hall, 1978.

Sparks, T. H., Greatorex-Davies, J. N., Mountford, J. O., Hall, M. L. & Marus, R. H. 'The effects of shade on the plant communities of rides in plantation woodland and implications for butterfly conservation'. *Forest Ecol. Management* **80**. 1996, 197–207.

Sparrow, H. R., Sisk, T. D., Ehrlich, P. R. & Murphy, D. D. 'Techniques and guidelines for monitoring Neotropical butterflies'. *Conserv. Biol.* **8**. 1994, 800–9.

Spitzer, K. 'Seasonality of the Butterfly fauna in Southeastern Vietnam (Papilionoidea)'. *J. Res. Lepid.* **22**. 1983, 126–30.

Spitzer, K., Jarôs, J., Havelka, J. & Leps, J. 'Effect of small-scale disturbance on butterfly communities of an Indochinese montane rainforest'. *Biol. Conserv.* **80**. 1997, 9–15.

Spitzer, K., Leps, J. & Soldan, T. 'Butterfly communities and habitat of seminatural savanna in southern Vietnam (Papilionoidea, Lepidoptera)'. *Acta Entomol. Bohemoslov.* **84**. 1987, 200–8.

Straatman, R. 'Summary of survey on ecology of *Ornithoptera alexandrae*' [Consultant report]. Papua New Guinea Department of Agriculture, Stock and Fisheries, 1970.

Stubbs, A. E. 'Is there a future for butterfly collecting in Britain?' *Proc. Trans. Brit. ent. nat. Hist. Soc.* **18**. 1985, 65–73.

Stubbs, A. E. 'A correction to Butterfly Conservation's claimed attitudes in invertebrate conservation'. *Brit. J. Ent. Nat. Hist.* **8**. 1995, 171–4.

Sunrose, T. 'Environmental evaluation for butterfly assemblages'. *Decline and Conservation of Butterflies in Japan*. **III**. 1996, 110–5.

Swengel, A. 'Effects of fire and hay management on abundance of prairie butterflies'. *Biol. Conserv.* **76**. 1996, 73–85.

Tabashnik, B. E., Perreira, W. D., Strazanac, J. S. & Montgomery, S. L. 'Population ecology of the Kamehameha butterfly (Lepidoptera: Nymphalidae'. *Ann. entomol. Soc. America* **85**, 1992, 282–5.

Tatham, J. T. 'The British Butterfly Conservation Society'. *Atala* 7. 1981, 53–4.

Taylor, R. W. 'A submission to the inquiry into the impact on the Australian environment of the current woodchip industry program'. Aust. Senate Hansard (Ref. Woodchip Inquiry), 12 August 1976, 3724–31.

Thomas, C. D. 'The status and conservation of the butterfly *Plebejus argus* L. (Lepidoptera: Lycaenidae) in north west Britain'. *Biol. Conserv.* **33**. 1985, 29–51.

Thomas, C. D. 'Ecology and conservation of butterfly metapopulations in the fragmented British Landscape'. *Ecology and conservation of butterflies*. Pullin, A. S. (ed.). London, Chapman & Hall, 1995, 46–63.

Thomas, C. D. & Jones, T. M. 'Partial recovery of a skipper butterfly (*Hesperia comma*) from population refuges: lessons for conservation in a fragmented landscape'. *J. Anim. Ecol.* **62**. 1993, 472–81.

Thomas, C. D. & Harrison, S. 'Spatial dynamics of a patchily distributed butterfly species'. *J. Anim. Ecol.* **61**. 1992, 437–46.

Thomas, C. D. & Mallorie, H. C. 'Rarity, species richness and conservation: butterflies of the Atlas Mountains in Morocco'. *Biol. Conserv.* **33**. 1985, 95–117.

Thomas, C. D., Thomas, J. A. & Warren, M. S. 'Distributions of occupied and vacant butterfly habitats in fragmented landscapes'. *Oecologia* **92**. 1992, 563–7.

Thomas, J. A. 'The ecology of the large blue butterfly'. *Rep. Inst. terr. Ecol.* (1976). 1977, 25–7.

Thomas, J. A. 'Why did the Large Blue become extinct in Britain?' *Oryx* **15**. 1980, 243–7.

Thomas, J. A. 'Butterfly year 1981–82'. *Atala* 7. 1981, 52–4.

Thomas, J. A. 'A quick method of estimating butterfly numbers during surveys'. *Biol. Conserv.* **16**. 1983a, 195–211.

Thomas, J. A. 'The ecology and conservation of *Lysandra bellargus* (Lepidoptera: Lycaenidae) in Britain'. *J. appl. Ecol.* **20**. 1983b, 59–83.

Thomas, J. A. 'A "WATCH" census of common British butterflies'. *J. Biol. Educ.* **17**. 1983c, 333–8.

Thomas, J. A. 'The conservation of butterflies in temperate countries: past efforts and lessons for the future'. *The Biology of Butterflies*. Vane-Wright, R. I. & Ackery, P. R. (eds). London, Academic Press, 1984, 333–53.

Thomas, J. A. 'Rare species conservation: case studies of European butterflies'. *The scientific management of temperate communities for conservation*. Spellerberg, I. F., Goldsmith, B. & Morris, M. G. (eds). Oxford: Blackwell, 1991, 149–97.

Thomas, J. A. 'Conserving European butterflies in modern landscapes'. *Decline and conservation of butterflies in Japan* **III**. 1996, 22–32.

Thomas, J. A. & Elmes, G. W. 'The ecology and conservation of *Maculinea* butterflies and their ichneumon parasites'. *Future of butterflies in Europe: strategies for survival*. Pavlicek-van Beek, T., Ovaa, A. H. & van der Made, J. G. (eds). Wageningen, Agricultural University, 1992, 116–23.

Thomas, J. A. & Simcox, D. J. 'A quick method for estimating larval populations of *Melitaea cinxia* L. during surveys'. *Biol. Conserv.* **22**. 1982, 315–22.

Thomas, J. A., Thomas, C. D., Simcox, D. J. & Clarke, R. T. 'Ecology and declining status of the silver-spotted skipper butterfly (*Hesperia comma*) in Britain'. *J. appl. Ecol.* **23**. 1986, 365–80.

Tinbergen, N. 'Ethologische Beobachtungen am Samtfalter, *Satyrus semele*'. *J. Orn.* **89**. 1941, 132–44.

Turner, J. R. G. 'Mimicry: the palatability spectrum and its consequences'. *The Biology of Butterflies*. Vane-Wright, R. I. & Ackery, P. R. (eds). London, Academic Press, 1984, 141–61.

Tyler, H., Brown, K. S. & Wilson, K. *Swallowtail butterflies of the Americas*. Gainesville, Scientific Publishers, 1994.

Urquhart, F. A. *The Monarch Butterfly*. Toronto, University of Toronto Press, 1960.

Urquhart, F. A. & Urquhart, N. R. 'Overwintering areas and migratory routes of the Monarch butterfly (*Danaus plexippus*, Lepidoptera: Danaidae) in North America, with special reference to the western population'. *Can. Ent.* **109**. 1977, 1583–9.

Usher, M. B. 'An assessment of conservation values within a Site of Special Scientific Interest in north Yorkshire'. *Field Stud.* 5. 1980, 323–48.

Usher, M. B. (ed.) *Wildlife conservation evaluation*. London, Chapman & Hall, 1986.

van der Heyden, T. 'Butterfly houses — a chance for the conservation of European butterflies, education and research?' *Future of butterflies in Europe*. Pavlicek-van Beek, T., Ovaa, A. H. & van der Made, J. G. (eds). Wageningen, Agricultural University. 1992, 315–8.

Vane-Wright, R. I. 'Ecological and behavioural origins of diversity in butterflies'. *Diversity in insect faunas*. Mound, L. A. & Waloff, N. (eds). Oxford, Blackwell, 1978, 56–70.

van Praagh, B. D. 'Adult and larval counts of the Eltham Copper Butterfly, *Paralucia pyrodiscus lucida* Crosby, 1993–1995'. Flora and Fauna Technical Report No. 144, Department of Natural Resources and Environment, Victoria, 1996.

van Swaay, C. A. M. 'An assessment of the changes in butterfly abundance in The Netherlands during the 20th century'. *Biol. Conserv.* **52**. 1990, 287–302.

van Swaay, C. A. M. 'The Dutch butterfly mapping scheme: methods and problems'. *Future of butterflies in Europe: strategies for survival*'. Pavlicek-van Beek, T., Ovaa, A. H. & van der Made, J. G. (eds) Wageningen, Agricultural University, 1992, 90–7.

van Swaay, C. A. M. 'Measuring changes in butterfly abundance in The Netherlands'. *Ecology and conservation of butterflies*. Pullin, A. S. (ed.). London, Chapman & Hall. 1995, 230–47.

Vasainanen, R. & Somerma, P. 'The status of *Parnassius mnemosyne* (Lepidoptera, Papilionidae) in Finland'. *Notul. Entomol.* **65**. 1985, 109–18.

Vaughan, P. J. 'The Eltham Copper Butterfly draft management plan'. Melbourne, *Arthur Rylah Institute for Environmental Research, Tech. Rpt* No. 57, 1987.

Vaughan, P. J. 'Management plan for the Eltham Copper butterfly (*Paralucia pyrodiscus lucida* Crosby) (Lepidoptera: Lycaenidae)'. Melbourne, *Arthur Rylah Institute for Environmental Research, Tech. Rpt* No. 79, 1988.

Vickery, M. 'Gardens: the neglected habitat'. *Ecology and conservation of butterflies*. Pullin, A. S. (ed.). London, Chapman & Hall. 1995, 123–34.

Vickery, M. 'The garden butterfly survey 1995'. *Butterfly Conservation News* **62**, 1996, 25–27.

Viejo, J. L. 'Diversity and species richness of butterflies and skippers in central Spain habitats'. *J. Res. Lepid.* **24**. 1985, 364–71.

Viejo, J. L. De Viedma, M. G. & Martinez Falero, E. 'The importance of woodlands in the conservation of butterflies (Lep.: Papilionoidea and Hesperioidea) in the centre of the Iberian Peninsula'. *Biol. Conserv.* **48**. 1989, 101–14.

Warren, M. S. 'The ecology of the wood white butterfly *Leptidea sinapis* (L.) (Lepidoptera: Pieridae)'. Ph.D. Thesis, University of Cambridge (not seen: referred to by Warren 1984), 1981.

Warren, M. S. 'The biology and status of the Wood White butterfly, *Leptidea sinapis* L. (Lepidoptera: Pieridae) in the British Isles'. *Ent. Gazette* **35**. 1984, 207–23.

Warren, M. S. Unpublished background papers on Butterfly Site Register. Nature Conservancy Council, Newbury, Berks, 1985a.

Warren, M. S. 'The influence of shade on butterfly numbers in woodland ridges, with special reference to the Wood White. *Leptidea sinapis*'. *Biol. Conserv.* **29**. 1985b, 287–305.

Warren, M. S. 'The ecology and conservation of the Heath Fritillary butterfly, *Mellicta athalia*. I. Host selection and phenology'. *J. appl. Ecol.* **24**. 1987a, 467–82.

Warren, M. S. 'The ecology and conservation of the Heath Fritillary butterfly, *Mellicta athalia*. II. Adult population structure and mobility'. *J. appl. Ecol.* **24**. 1987b, 483–98.

Warren, M. S. 'The ecology and conservation of the Heath Fritillary butterfly, *Mellicta athalia*. III. Population dynamics and the effect of habitat management'. *J. appl. Ecol.* **24**. 1987c, 499–513.

Warren, M. S. 'The successful conservation of an endangered species, the heath fritillary butterfly, *Mellicta athalia*, in Britain'. *Biol. Conserv.* **55**. 1991, 37–56.

Warren, M. S. 'The conservation of British butterflies'. *The ecology of butterflies in Britain*. Dennis, R. L. H. (ed.). Oxford, Oxford University Press. 1992, 246–74.

Warren, M. S. 'A review of butterfly conservation in central southern Britain: I. Protection, evaluation and extinction on prime sites'. *Biol. Conserv.* **64**. 1993a, 25–35.

Warren, M. S. 'A review of butterfly conservation in central southern Britain: II. Site management and habitat selection of key species'. *Biol. Conserv.* **64**. 1993b, 37–49.

Warren, M. S. 'The UK status and metapopulation structure of a threatened European butterfly, *Eurodryas aurinia* (the marsh fritillary)'. *Biol. Conserv.* **67**. 1994, 239–49.

Warren, M. S., Thomas, C. D. & Thomas, J. D. 'The status of the Heath Fritillary butterfly, *Mellicta athalia* Rott., in Britain'. *Biol. Conserv.* **29**. 1984, 287–305.

Warren, M. S., Pollard, E. & Bibby, T. J. 'Annual and long-term changes in a population of the wood white butterfly *Leptidea sinapis*'. *J. Anim. Ecol.* **55**. 1986, 707–20.

Waterhouse, G. A. 'The genus *Heteronympha* in New South Wales'. *Proc. Linn, Soc. N.S.W.* 1897, 240–3.

Waterhouse, G. A. 'A second monograph of the genus *Tisiphone* Hubner'. *Aust. Zool.* **5**. 1928, 217–40.

Waterhouse, G. A. *What Butterfly is That?* Sydney, Angus & Robertson, 1932.

Waterhouse, G. A. & Lyell, G. *The Butterflies of Australia*. Sydney, Angus & Robertson, 1914.

Watt, W. B., Chew, F. S., Snyder, L. R. G., Watt, A. G. & Rothschild, D. E. 'Population structure of pierid butterflies. I. Numbers and movements of some montane *Colias* species'. *Oecologia* **27**. 1977. 1–22.

Wellington, W. G. 'A special light to steer by' *Nat. Hist.* **83**. 1974, 46–53.

Wells, S. M., Pyle, R. M. & Collins, N. M. *The IUCN Invertebrate Red Data Book*. Gland and Cambridge, IUCN, 1983.

Whalley, P. E. S. *Butterfly Watching*. London, Severn House, 1980.

White, R. R. 'Inter-peak dispersal in alpine checkerspot butterflies (Nymphalidae)'. *J. Lepidopt. Soc.* **34**. 1980, 353–62.

Wickler, W. *Mimicry in plants and animals*. London, Weidenfeld & Nicolson, 1968.

Wiklund, C. & Fagerstrom, T. 'Why do males emerge before females? A hypothesis to explain the incidence of protandry in butterflies'. *Oecologia* **31**. 1977, 153–8.

Wiklund, C. & Forsberg, J. 'Courtship and male discrimination between virgin and mated females in the orange tip butterfly *Anthocaris cardamines*'. *Anim. Behav.* **34**. 1985, 328–32.

Wiklund, C., Persson, A. & Wickman, P. O. 'Larval aestivation and direct development as alternative strategies in the speckled wood butterfly, *Pararge aegeria*, in Sweden'. *Ecol. Entomol.* **8**. 1983, 233–8.

Wilcove, D. S., McMillan, M. & Winston, K. C. 'What exactly is an Endangered Species? An analysis of the U.S. Endangered Species List: 1985–1991'. *Conserv. Biol.* **7**. 1993, 87–93.

Wilcox, B. A., Murphy, D. D., Ehrlich, P. R. & Austin, G. T. 'Insular biogeography of the montane butterfly faunas in the Great Basin: comparison with birds and mammals'. *Oecologia* **69**. 1986, 188–94.

Williams, C. B. *The Migration of Butterflies*. Edinburgh, Oliver & Boyd, 1930.

Williams, C. B. *Insect Migration*. London, Collins, 1958.

Williams, E. H. 'Fire-burned habitat and reintroductions of the butterfly *Euphydryas gillettii* (Nymphalidae). *J. Lepidopt. Soc.* **49**. 1995, 183–91.

Wood, P. & Samways, M. J. 'Landscape element pattern and continuity of butterfly flight paths in an ecologically landscaped botanic garden, Natal, South Africa'. *Biol. Conserv.* **58**. 1991, 149–66.

Xerces Society. *Butterfly Gardening*. Portland, Oregon, Xerces Society, 1990.

Yamamoto, M. 'Notes on the methods of belt transect census of butterflies'. *J. Fac. Sci. Hokkaido Univ., Zoology* **20**. 1975, 93–116.

Yamamoto, M. 'A comparison of butterflies assemblages in and near Sapporo city, northern Japan'. *J. Fac. Sci. Hokkaido Univ., Zoology* **20**. 1977, 621–46.

Young, A. M. 'Evolutionary responses by butterflies to patchy spatial distributions of resources in tropical environments'. *Acta Biotheoret.* **29**. 1980, 37–64.

Zalucki, M. P. 'The monarch butterfly — a non-pest exotic insect'. *The ecology of exotic animals and plants. Some Australian case histories*. Kitching, R. L. (ed.) Brisbane, John Wiley, 1986.

Zeuner, F. E. (1962) Notes on the evolution of the Rhopalocera (Lepidoptera). *Proc. XIth Int. Congr. Entomol.* **1** (1960). 1962, 310–13.

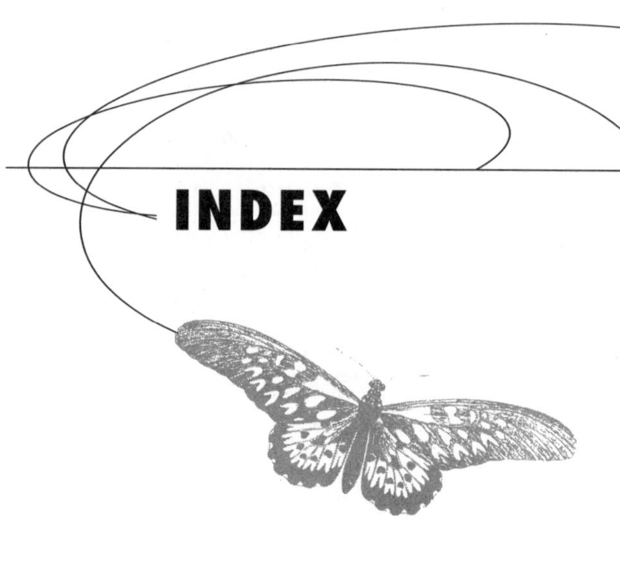

INDEX